AF348530

A Universe of Earths

A Universe of Earths

Our Planet and Other Worlds, from Copernicus to NASA

DENNIS DANIELSON
CHRISTOPHER M. GRANEY

OXFORD
UNIVERSITY PRESS

OXFORD

UNIVERSITY PRESS

Oxford University Press is a department of the University of Oxford.
It furthers the University's objective of excellence in research, scholarship,
and education by publishing worldwide. Oxford is a registered trade mark of
Oxford University Press in the UK and in certain other countries.

Published in the United States of America by Oxford University Press
198 Madison Avenue, New York, NY 10016, United States of America.

CIP data is on file at the Library of Congress

ISBN 9780197803516

DOI: 10.1093/9780197803547.001.0001

The manufacturer's authorized representative in the EU for product safety is
Oxford University Press España S.A. of Parque Empresarial San Fernando de Henares,
Avenida de Castilla, 2 – 28830 Madrid (www.oup.es/en or product.safety@oup.com).
OUP España S.A. also acts as importer into Spain of products made by the manufacturer.

For my grandsons: Francis, Ivan, Zekie, and Billy

*For my sister, Laura Kathleen Graney (1971–2013), who would have loved
and hated this book.*

Contents

List of illustrations

Preface

"The most beautiful star in the heavens"

Planet Earth. The phrase trips lightly from our tongues. Yet *planet* Earth has been a concept for a mere fraction of our recorded history. Even only four centuries ago, when Nicolaus Copernicus's model of a Sun-centered world was just starting to gain acceptance, the concept of Earth as a "wandering star"—the core sense of "planet," from Greek *planēs*, "wanderer"—still seemed to most people strange and contradictory.

Today, we know that when we stand on the seashore gazing *out* at a sunrise or sunset, we're actually moving and looking *in* toward the center of the Solar System, with the orbits of the inferior planets (Mercury and Venus) between us and the Sun, and those of the superior planets (Mars, Jupiter, Saturn, Uranus, and Neptune) behind us. We also know that, from space, Earth is *visible.* In the words of Apollo astronaut Gene Cernan, who saw it with his own eyes, even from "a quarter of a million miles away, [you're] looking at the most beautiful star in the heavens."

Until about the mid-1600s, however, most humans thought of Earth as, well, just Earth. It was fixed beneath our feet. It was nothing like certain lights in the heavens, the planets, which did not remain fixed in a constellation with the other stars but rather roved over time from one constellation of the "Zodiac" (Leo, Virgo, etc.) to another and bore names like "Venus" and "Mars." Unlike them, Earth was excluded from what Galileo called "the dance of the stars." Earth and the planets were even thought to be made from different kinds of stuff: The heavenly bodies were light, quintessential, ethereal; Earth was dirt. Artists depicted it as dim, nonluminous, scarcely visible from a cosmic perspective.

In this book we trace the story of how that all changed. We unfold how the Copernican Revolution's "starring" of Earth served to augment and enrich our understanding of where Earth and its inhabitants are in the Universe, how we fit into the Big Picture of the Cosmos. And of course that's an understanding toward which astronomers, cosmologists, and many other curious minds are still striving.

Furthermore, Earth's upgraded planetary status gave birth to an intriguing and powerful analogy: If Earth *is* a planet, then surely the planets might be other earths. This bold conjecture ignited a vast, lively literature of space travel, extraterrestrials, and other worlds. Still more importantly, it provided the impetus for scientific research, research that continues to flourish today. Ever since Copernicus (1473–1543) and his followers proposed that our world shares something in common with those roving lights in the night sky, namely, motion around the Sun, the possibility of other earths populated by other intelligent beings—a hypothesis that came to be known as "Plurality of Worlds"—has fascinated scientists and adventurers alike. Without that analogy between Earth and other planets, how could we ever have imagined the worlds of *Avatar, Star Wars,* and the Marvel Universe?

But the hypothesis of a Cosmos full of other earths, for all its appeal—and for all its value in stimulating both scientific and popular imaginations—has so far been uncorroborated by observations of the actual Universe. In fact, the reality of science began undermining that hypothesis right from the start. No less a figure than the great Copernican astronomer Johannes Kepler (1571–1630) explained how scientific reasoning and measurement militated against the notion of other earths beyond our Solar System.

Nonetheless, Plurality of Worlds has endured. It has given rise not only to space-based science fiction but also to a panoply of theories about alien beings and their activities, from the weird stuff scientists generally dismiss as fiction—UFO (or UAP) mania; conspiracy theories involving lizard people and the like; and claims of alien abductions—to other, more serious claims: for example, that we can best explain some astronomical phenomena by invoking the existence of powerful extraterrestrial civilizations. Plurality of Worlds has also generated finger-wagging about our "insignificant" place in the Universe, about Copernicus demoting Earth, "dethroning" it—and us—from our supposedly once-privileged central location. This skewed interpretation of the last half millennium of astronomical history offers a "Copernican" process in which science renders Earth cosmologically meaningless—just an ordinary planet orbiting an ordinary star in an ordinary galactic location.

By contrast, the surprising story we trace in this book—surprising in the eyes of its original characters and perhaps surprising for us today—challenges Plurality and its version of history. For what science truly offers is a different account, one that releases us humans from the pre-Copernican view of Earth as low, lowly, dark, a cosmic sump, something confined within (as one writer put it in 1486) the "excrementary and filthy parts of the lower world." Instead, from the day Copernicus first drew that circle depicting Earth's Sun-centered orbit, we have been offered the bracing realization that Earth is, in the classical sense, a star, a dynamically wandering one, and a shining participant in the dance of the stars. The last half millennium of history offers a story *starring the Earth* in more than one sense of the phrase.

What follows, then, is a reframing of some of the most prominent, profound, and exhilarating insights that continue to shed light on Earth's place, our place, in the Universe—insights woven from eureka moments, from rigorous scientific scrutiny, and from the history that we encounter on the road from Copernicus to NASA and the present.

Note on text and usage

We follow the style of the International Astronomical Union in capitalizing Earth, Sun, Moon, Galaxy, Universe, Cosmos, and the like, when these designate a single astronomical entity (parallel to the standard usage whereby we capitalize Mercury, Venus, and the proper names of other planets). When a word is generic or plural, however, it is not capitalized (e.g., "Galileo discovered four moons of Jupiter," "Humans are created from earth," etc.).

We use American punctuation and spelling except for the word "storey." It is important to our argument that, when we speak, for example, of the upper and lower *storeys* of the pre-Copernican Universe, these architectural/cosmographical designations should not be confused with narratives or stories.

In quotations of primary material, we have taken the liberty of modernizing English spelling—which we would *not* have done if this were a work of textual literary criticism.

In citations of writings in which no published translation is indicated, the translation is our own. For some lesser-known works, we occasionally provide the original in the footnotes or in parentheses. For better-known authors whose original works are readily available (e.g., Copernicus, Galileo, Kepler), we quote either our own or a standard published translation.

Footnotes are presented in the shortest form possible. Full publication details for the references are given at the end of the book in Works Cited. The following are three frequently cited sources, abbreviated thus:

BOTC. The Book of the Cosmos: Imagining the Universe from Heraclitus to Hawking. Edited by D. Danielson. 2000; Basic Books, 2001.

KCGSM. Kepler, Johannes. *Kepler's Conversation with Galileo's Sidereal Messenger.* Translated by Edward Rosen. Johnson Reprint Corporation, 1965.

MathDisq: Johann Georg Locher with Christoph Scheiner. *Disquisitiones Mathematicae.* Ingolstadt, 1614; translated in Christopher M. Graney. *Mathematical Disquisitions: The Booklet of Theses Immortalized by Galileo.* University of Notre Dame Press, 2017.

1

Prologue

Discerning the Cosmos

Astronomer Virginia Trimble tells the story of how in 1968, as a 24-year-old graduate student passing through London, she was invited by Frank Drake (1930–2022), founder of SETI (the scientific <u>S</u>earch for <u>ExtraT</u>errestrial <u>I</u>ntelligence) to accompany him to the British Museum to "meet" the Rosetta Stone, the "stele" that more than a century earlier had permitted scholars to decipher ancient Egyptian hieroglyphics. Drake told Trimble how in the early 1950s, during the Korean War, he had served in the military as a codebreaker. This led him to reflect that perhaps the most interesting, challenging message to decipher might be one written by a different species, from a different planet, perhaps one orbiting a nearby Sun-like star such as Epsilon Eridani or Tau Ceti.[1]

Drake's thirst for communication with the "other" and the "elsewhere" had motivated him seven years earlier, in 1961, to organize the first concerted scientific effort to conduct a search for life "out there" in our Galaxy and perhaps beyond. At the Green Bank Radio Observatory in West Virginia, Drake assembled a team of a dozen distinguished scientists from across the disciplines—including men like Carl Sagan (then twenty-seven) and Nobel biochemist Melvin Calvin—to pursue "Project Ozma": to take first steps toward establishing radio contact with, or receiving it from, intelligent civilizations beyond Earth. These efforts eventually grew into the SETI Institute, where today, more than six decades later, a cadre of scientists pursue various aspects of Drake's quest, while numerous researchers at other institutions pursue related fields like astrobiology.

We'll return in later chapters to SETI and the wider quest for other worlds. But because this book is also concerned with some of the human factors that

---

[1] Anecdote based on a telephone interview kindly granted by Virginia Trimble, November 11, 2022.

A Universe of Earths. Dennis Danielson and Christopher M. Graney, Oxford University Press.
© Oxford University Press (2025). DOI: 10.1093/9780197803547.003.0001

shape and inspire science here on Earth, let's ponder one further anecdote that may illuminate a major element of Drake's search.

In the early 1970s, more than a decade after the inception of Project Ozma, budding science writer Dava Sobel, then in her twenties, attended an introductory training session offered by Tompkins County Suicide Prevention in Ithaca, New York, home of Cornell University. Sobel had had enough experience with people contemplating suicide that she wanted to learn more about intervention. Also attending that meeting, she recounts, was a fortyish man with pure white hair, wearing a dark shirt. Sobel recalls somehow wondering if he might be a janitor. But during the round of self-introductions, the man said, "My name's Frank and I'm an astronomer." He explained that in his line of work, conducted mainly at night and in remote places, people are far away from their families. "They're lonely and under a lot of stress, and anything can happen. . . . Once, I took a loaded gun out of someone's hand."[2] Concerned that this might happen again, he wanted to acquire skills for dealing with such situations.

Frank Drake completed the training and worked the all-night shift at Suicide Prevention as a crisis intervention counselor for one or two Fridays a month through that decade. A passionate, committed valuing of human, terrestrial life—and of listening intently and discerningly—was part of the makeup of the man who also, more than anyone else, spurred the search for life elsewhere in the Galaxy.

. . .

In this book, we too want to listen intently to those who have led the way in discerning the character of our Cosmos and understanding our place and Earth's place within it. You could say that both Nicolaus Copernicus, who died in 1543, and Frank Drake, who died in September 2022, were remarkable for their humane, life-affirming, value-driven commitment to "listen" to the Universe and to probe, and try to grasp, its character. To listen well, both for them and for us, often demands a suspension of what we thought we already knew about people, about the world, and in some cases about history. It requires a willingness to have our assumptions and expectations surprised.

Such listening may even cause us to rethink our understanding of the nature of science. To assess what counted as science from the era of

---

[2] Anecdote based on a telephone interview kindly granted by Dava Sobel, December 16, 2022. See also Drake and Sobel, *Is Anyone Out There?*, 214.

Copernicus to that of Frank Drake, we need to comprehend the scientific worlds within which such figures worked. Their soaring imaginations and curiosity were admirable and necessary, but not sufficient. The channels in which their scientific minds ran were constrained by demands for coherence, mathematical consistency, and rigorous attention to the evidence available *to them in their times*. Mathematics may touch the eternal, to paraphrase the distinguished mathematician Sing-Tung Yau,[3] but science changes and grows as our access to evidence changes and grows. Drake could therefore use the same Pythagorean Theorem known to Copernicus, but he would have found most of Copernicus's science obsolete.

That is why Drake endorsed the idea of a "living science" that nonetheless "thrives on skepticism."[4] It is also why SETI has almost nothing to do with "close encounters," "little green men," or other matters related to "ufology." Moreover, note that Drake was a radio astronomer and that SETI started with radio. It's not just on Earth where radio has advantages as a means of communication. This makes SETI a quite recent endeavor: That our broadcasting of radio waves started just a little more than a century ago implies that Earth's "radio bubble" now forms a sphere that is only about two hundred light-years in diameter—this in a Galaxy approaching 100,000 light-years across.

But speculation about extraterrestrial intelligence (ETI) is not recent. It predates the twentieth century and even precedes the time of Copernicus, although the Polish astronomer's achievements energized it immeasurably. Johannes Kepler, a brilliant mathematician and one of only a handful of serious scientists to adopt Copernicus's Sun-centered, or *heliocentric*, model of the Universe before the year 1600, most famously worked out the true mathematical nature of how planets trace elliptical orbits. However, Kepler also felt right at home with the idea of ETI. Responding to Galileo's 1610 account of the moons of Jupiter recently discovered with his telescope, he expressed delight that these moons would illuminate the skies of that stupendous planet. Mind you, Kepler betrayed a bit of terrestrial chauvinism by declaring Earth's place in what we now call the Solar System to be "central" (with Mars, Jupiter, and Saturn on the starward side and the Sun, Mercury, and Venus on the other side). He speculated that Jupiter's inhabitants nevertheless enjoyed *four* moons as against our one moon as a kind of divine compensation for their less-than-ideal location: "Let the

---

[3] Nadis and Yau, *The Gravity of Math*, 6, 216.
[4] Drake and Sobel, *Is Anyone Out There?*, 63, 208.

Jovian creatures, therefore, have something with which to console themselves. Let them . . . have their own planets [i.e., their moons]. We humans who inhabit the earth can with good reason (in my view) feel proud of the pre-eminent lodging place of our bodies, and we should be grateful to God the creator."[5]

But no one today associates ETI with Jupiter. Today's science reveals that the Jovian planet lacks even a surface to stand on, let alone inhabitants standing there, gazing at its moons. It is a ball of turbulent gas. Its winds far surpass Earth's fiercest hurricanes. Beneath its colorful clouds that we see with our telescopes and space probes, the density of that gas just increases with depth. Were you somehow, right now, teleported to any spot on (or in) Jupiter, you would die—quickly.

That is true for pretty much any place in the Solar System outside of Earth. When we think today of extraterrestrials, we don't think of them coming from one of the planets that orbit the Sun. Invading Jovians, or better, Martians, are a late-nineteenth- through mid-twentieth-century idea. The UFOs in H. G. Wells's 1898 book *War of the Worlds* came from Mars; likewise, those in Orson Welles's 1938 radio version of Wells's story, and in the 1953 movie version. In Steven Spielberg's 2005 *War of the Worlds* movie, however, Mars was not part of the story. Today we imagine ETI coming from other suns—from the stars. It is the stars that SETI looks to, not Jupiter or Mars.

Kepler would have had none of that. He had no tolerance for the idea of the stars being other suns with planets orbiting them. He believed that there was one Sun in the Universe: ours. Correction—he did not *believe* that; he was scientifically *certain* of it. The simplest observations, measurements, and calculations proved it. Just look at the sky, he said in that 1610 response to Galileo:

> If [the fixed stars] are suns having the same nature as our Sun, why do not these suns collectively outdistance our Sun in brilliance? Why do they all together transmit so dim a light . . . ? When sunlight bursts into a sealed room through a hole made with a tiny pinpoint, it outshines the fixed stars at once. The difference is practically infinite.[6]

In other words, stars look nothing like the Sun. Kepler took into account the weak light output of the stars, their apparent sizes in the sky, and their

---

[5] *KCGSM* [Kepler, *Kepler's Conversation with Galileo's Sidereal Messenger*, tr. Rosen], 46.
[6] *KCGSM*, 35.

vast distances from Earth as required by Copernicus. Using basic geometry, he showed that stars were huge and that stars were dim—again, nothing like the Sun. In 1604 Kepler published a calculation showing that Sirius (the Dog Star, the most prominent star visible in the night sky) had to be larger than the orbit of Saturn. Every last star we can see, even those barely visible to the eye, had to be larger than Earth's orbit. The Sun was nothing in size compared to the smallest visible star.

But size did not matter, as Kepler saw it; what mattered were energy and life. In terms of light output, he said, the tiny Sun surpasses all the giant stars—combined—by a practically infinite amount.

Earth is tinier still, Kepler said,

and yet behold . . . our little ball, the little cottage of us all, which we call the Earth: the womb of the growing, herself informed by a certain internal faculty. The architect of marvelous works, she kindles daily so many little living things from herself—plants, fishes, insects—as she may easily scorn the rest of the bulk in view of this her nobility.

And even tinier than Earth are

these fine bits of dust called human beings; to whom the Creator has granted such, that in a certain way they may beget themselves, clothe themselves, arm themselves, teach themselves infinite arts, and daily work toward the better; in whom is the image of God; who are, in a certain way, lords of the whole bulk.

And who among us would choose a body the breadth of the universe in exchange for no soul? Let us learn, therefore, what well pleases the Creator, who is the author of both coarse bulks and minute perfections. For he glories not in bulk, but ennobles those he willed to be small.[7]

Kepler promoted his "single Sun" Universe in the process of explicitly criticizing the ideas of Giordano Bruno (1548–1600). Bruno had urged that the stars were other suns, which would be orbited by other earths, which would be inhabited. Today, we envision ETI sending out radio signals from those sorts of places: other earths around other suns. We envision a Universe like Bruno's, not Kepler's.

Bruno had been burned at the stake in 1600. He was executed partly, perhaps, for his insistent advocacy of countless other suns and earths. It has

---

[7] Kepler, "Chapter 16 of *De Stella Nova*," 58.

become fashionable to treat Bruno as a martyr to science, as a man who, as historian David Wootton puts it, "on crucial points . . . was right before anyone else."[8] Bruno was in fact burned mainly for his advocacy of theological heresies that his contemporaries found deeply offensive.[9] But quite apart from theology, Kepler wanted to make it clear that Bruno's ideas about the physical Universe had no basis in science. Any astronomer who could see the sky, make basic measurements, and do some simple math could reproduce Kepler's results and see that there was just one Sun and that Bruno was wrong about Plurality of Worlds.

Yet if science showed that there was only one Sun, how did Bruno's idea that there were other suns, with other earths orbiting them, ever catch on? How did he come to be viewed as "right" and as a martyr to science? That is a big part of our surprising story. It will take up a good bit of this book. But as a hint, consider something else about Kepler.

He wrote in his 1618 *Epitome of Copernican Astronomy* that it would seem fitting and logical if the sizes of planets increased with their distance from the Sun: "nothing is more consistent with nature than that the order of their magnitudes should be the same as the order of their spheres, so that of the six primary planets, Mercury should be the least, Saturn the greatest, inasmuch as the former is moved in the smallest, the latter in the largest orbit."[10] Others liked this idea too.

For example, a few decades after *Epitome*, Jeremiah Horrocks (1618–1641), a young English astronomer, argued that planet diameters should scale with the diameters of their orbits; in this way, all the planets would appear to be the same size as seen from the Sun, including Earth. "Surely," he said, "if the Earth agrees with the others as to motion, and if the proportion of its orbit to that of the rest [following Kepler's laws of planetary motion] be so exact, it would be ridiculous that it should differ so markedly from the others in the proportion of its diameter."[11] Earth had to fit in.

Otto von Guericke agreed. In his 1672 *New Magdeburg Experiments*, he included a diagram of the Copernican sun and planets set in a Bruno-style Universe of sun-like stars that extended indefinitely out into space. He included in the diagram some things visible only telescopically, like

---

[8] Wootton, *The Invention of Science*, 149.

[9] Shackleford, "Myth 7. That Giordano Bruno Was the First Martyr of Modern Science," 59–67.

[10] Horrocks, *Venus Seen on the Sun*, 66. See Kepler, *Epitome of Copernican Astronomy*, 38–39, which Horrocks is citing.

[11] Horrocks, *Venus Seen on the Sun*, 69–71.

the moons of Jupiter, and the moon Titan orbiting Saturn. And he showed the planets increasing in size exactly as Horrocks said: Mars is bigger than Earth; Earth is bigger than Venus (see Figure 1.1).

But there was no evidence for such an arrangement. Earth and Venus are both larger than Mars. A decent telescope, even a decent one in Guericke's time, would show his diagram to be wrong. For example, if Mars and Jupiter were to appear to be the same size as seen from the Sun, as in the diagram, Mars would have to best Jupiter in size when seen from Earth. It doesn't (it is not even close). Again, any astronomer who could see the sky, make basic measurements, and do some simple math could see that this whole "increasing in size" proposal was wrong.

In 1698 the Dutch astronomer Christiaan Huygens (who discovered Titan) published the diagram seen in Figure 1.2 in his book *Celestial Worlds Discover'd*. He made the measurements, did the math, and got the sizes correct. But Guericke's diagram warns us that appealing ideas in science can catch on and stick around even when science itself does not support them. Huygens himself expressed surprise that planet sizes did not follow a seemingly logical progression: "'Tis remarkable," he said, "that the bodies of the

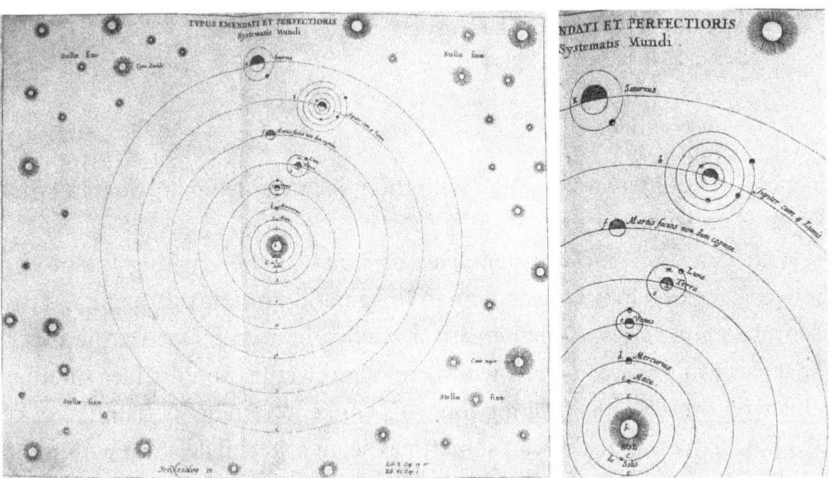

Figure 1.1 Left: Otto von Guericke's diagram of the Copernican Sun and planets amid a Universe of Stars. Right: Detail showing the planets increasing in size with distance from the Sun.
Credit: Smithsonian Libraries

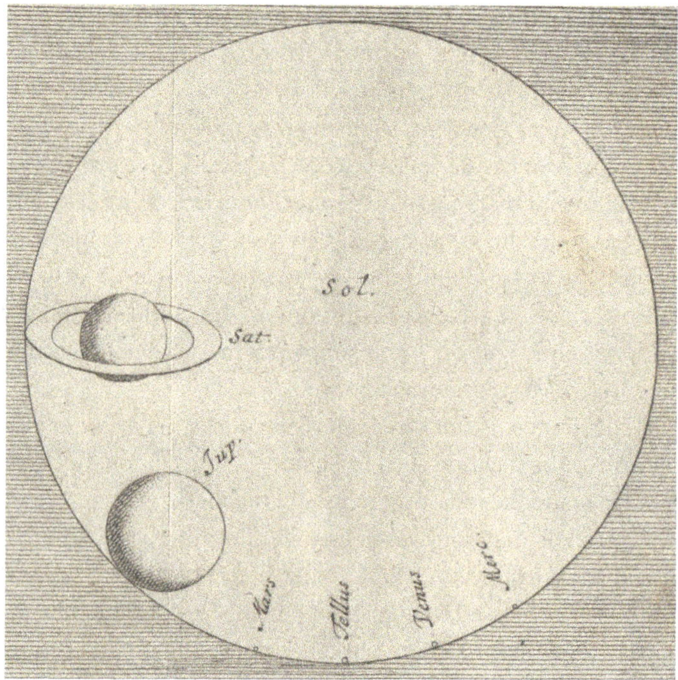

Figure 1.2 The relative sizes of the Sun and the planets as illustrated by Christiaan Huygens. The sizes are in general agreement with modern measurements.
Credit: ETH-Bibliothek Zürich

planets do not increase together with their distances from the Sun, but that Venus is much bigger than Mars."[12]

With Guericke and Huygens we see hints of the ongoing tug-of-war between the eager and elastic human imagination on the one hand and the rigorous empirical and mathematical science of a given age on the other. That same tug-of-war is seen with Bruno and Kepler and the idea of other suns with other earths orbiting them—a concept at the very foundation of SETI. And the tug-of-war can sometimes seem a stalemate, even within the mind and work of a single scientist. Thus, Frank Drake could aver that "science thrives on skepticism" but also assert that, "at this very minute, *with*

---

[12] Huygens, *Celestial Worlds Discover'd*, 17.

almost absolute certainty, radio waves sent forth by other intelligent civilizations are falling on the earth."[13] If such radio waves were discerned, then of course that would indeed establish the truth of the greatest of SETI's premises. Once again, any astronomer who could see the sky and make the measurements could see that.

Yet in fact—and for Drake, as we shall see, *surprisingly*—to date no such radio waves have been detected. Admittedly, none of that "absence of evidence" renders the quest vain. It certainly doesn't constitute "evidence of absence."[14] And yet bluntly and *scientifically*, absence of evidence *is* absence of evidence.

Clearly, much remains to be learned. For Copernicus too, much remained to be learned before his account of what we now call the Solar System could be verified. When it was, further lessons and discoveries still lay ahead. Much expansion and adjustment of our understanding of Earth as a planet would be demanded by the evidence. In the coming chapters, we'll trace just some of those significant expansions of knowledge about our Cosmos and in particular about the place and role of Earth, our own wandering star, within it. In the process we'll taste the difficulty, the exhilaration, and the surprises that attend the very human exercise of science upon which that knowledge is grounded.

---

[13] Frank Drake, "Who Led Search for Life on Other Planets, Dies at 92."
[14] The truism is reliably attributed to Astronomer Royal Martin Rees but was taken up by Carl Sagan, e.g., in *The Demon Haunted World*, 213.

2

Copernicus's Giant Leap

On July 20, 1969, as he first set foot onto the surface of the Moon, Apollo 11 astronaut Neil Armstrong spoke what would become his most famous utterance: "That's one small step for a man, one giant leap for mankind." Only a few years later, in 1973, Armstrong sent a picture of our blue-green world to Nicolaus Copernicus's alma mater, the Collegium Maius in Krakow, Poland. The photograph, which still hangs on the wall in the Collegium, bears the following handwritten inscription:

To the Copernicus Museum
     Krakow
On the 500th birthday
     Of a Giant
          Neil Armstrong
           Apollo 11

The Apollo photos of Earth taken from afar continue to evoke wonder. Their vivid message that we live on a beautiful world, a bright oasis in the blackness of space, can still take our breath away, even after more than a half century. And Armstrong recognized that Apollo's startling polychromatic vision of Earth, along with his own giant step onto another world on our behalf, was prepared for by the giant of an astronomer who first proposed that Earth is in fact a *planet*, a wandering star.

How did Copernicus himself (1473–1543), all those years ago, take his own first step along the planetary path? After all, for almost two millennia, Earth had been thought of as being located in the "downstairs" of a two-storey Universe, fixed at its center. That downstairs was the home of the dirty, the heavy, the sluggish, and the always-changing, home of things like rocks and mud—the stuff of our experience. The Sun was an eternal light in the sky, "upstairs," where nothing changes and things are light, perfect, and easy moving. Over the course of a year the Sun wandered through the constellations of the Zodiac. Yes, the Sun was a planet too, circling about the center.

A Universe of Earths. Dennis Danielson and Christopher M. Graney, Oxford University Press.
© Oxford University Press (2025). DOI: 10.1093/9780197803547.003.0002

This fit with common sense. Consider that even today we continue to check our weather apps or websites to discover when the Sun will *rise* and *set* tomorrow. We know that, whatever change befalls us here on Earth, tomorrow the Sun *will* rise, as it always does. And absent earthquakes, Earth *feels* fixed and unmovable.

Yet all was not well with the classical model of the Universe. The second-century Alexandrian astronomer Claudius Ptolemy had developed a magnificent astronomical model upon the ideas of the Greek philosopher Aristotle. Ptolemy's work was usually called *Almagest*, an Arabic title meaning "the greatest." Ptolemy and his followers, however, had to tinker with their model to accommodate observational anomalies that seemed to clash with the sorts of purely circular and regular motions that Aristotle had ascribed to the heavenly bodies upstairs. Perhaps best known of these was the phenomenon of "retrograde motion," most dramatically demonstrated by Mars. Every couple of years, whenever Mars is opposite the Sun in the sky, Mars appears to cycle back on itself as it moves through the constellations of the night sky, brightening as it does so, before resuming its path through the stars—as illustrated in this time-lapse photo of Mars wandering through the constellations over a period of weeks (see Figure 2.1).

Figure 2.1 Time-lapse photo of Mars wandering through the constellations over a period of weeks.
Credit: Rob Kerby Guevarra/IAU OAE

The device proposed to solve this anomaly was an *epicycle*: Mars was carried about on a smaller rotating circle (the epicycle) that was in turn borne along on a larger circle (the deferent) around the Earth. The double circling sufficed to explain both Mars' backing up and its brightening, but not its being opposite the Sun when these things happened.

There were other anomalies, and other devices were proposed to solve them. For example, to get Ptolemy's theory to line up better with Mars' observed overall motion, Mars' deferent circle had to be *eccentric*, off center from the center of the Universe (Earth's location). And even with all of Ptolemy's devices, he had no explanation for that important fact about the retrograde motion of Mars through the constellations: It was linked to the yearly motion of the Sun through the constellations, such that retrograde always occurred when Mars was opposite the Sun in the sky.

Over time the whole Ptolemaic system came to appear, in the eyes of critics—justifiably or not—as a sort of inelegant Rube Goldberg cosmic contraption. In the seventeenth century, the poet John Milton would satirize how astronomers

> wield
> The mighty frame; how build, unbuild, contrive
> To save appearances, how gird the sphere
> With centric and eccentric scribbled o'er,
> Cycle and epicycle, orb in orb.
>
> (*Paradise Lost* 8.80–84)

To tackle these problems, however, was no easy task. We don't know exactly how or when Copernicus's dissatisfaction with the extended Ptolemaic model arose. For one thing, he had much other work to do. He was born in Torun, Poland, and after receiving the best education that money and influence could buy—in Krakow, Bologna, and Padua, in medicine and canon law—he returned to the cathedral in Frombork on the Baltic coast, in the episcopal province of Varmia. There he served as a "canon," a sort of ecclesiastical civil servant and administrator (he was not a priest, as is sometimes wrongly claimed). Beyond these duties, the polymathic Copernicus was also valued as a doctor and was even recruited to write reports on the functioning of the local currency in Prussia.

Technically, then, as an astronomer, Copernicus was an amateur. Unlike his most famous successors—Tycho Brahe, Kepler, Galileo,

Newton—Copernicus was never paid for his science nor was he ever granted any patronage to help him pursue it. But his dedication to astronomy was all the more remarkable for his having so little support as well as essentially no audience for his ideas (or so he thought)—not to mention that he lived away in the hinterlands of European thought and culture, in a place he later called "this remotest corner of the Earth."

Nonetheless, at some point after 1510, as he approached his fortieth birthday, Copernicus wrote a short document—untitled, unsigned, and circulated privately—that from its opening page exposed its readers to a counterintuitive, even shocking new way of conceiving the Universe. This work, now known as the *Commentariolus* ("short commentary"), was circulated a bit like a "preprint" today that a scientist posts online in hopes of receiving some helpful response before the paper goes into publication. Yet the copies of the *Commentariolus* seem to have met with what Alfred Romer calls a "tantalizing silence."[1] In any case, the substance of the manuscript spelled such a radical departure from millennia-old assumptions and habits of mind that who could expect them to gain acceptance without a monumental mathematical and observational defense?

Precisely because he is about to introduce an astonishing set of ideas, Copernicus writes the *Commentariolus* in an almost flat, deadpan, "lab report" sort of style. Ptolemy's "planetary theories," he concedes, are "consistent with the numerical data." And yet, unfortunately, they "present no small difficulty"—he lists some of the problems with the "devices" just sketched. And just as importantly, they aren't "sufficiently pleasing to the mind."

> Having become aware of these defects, I often considered whether there could perhaps be found a more reasonable arrangement of circles, from which every apparent inequality would be derived and in which everything would move uniformly about its proper center, as the rule of absolute motion requires. After I had addressed myself to this very difficult and almost insoluble problem, the suggestion at length came to me how it could be solved with fewer and much simpler constructions than were formerly used, if some assumptions . . . were granted me.[2]

---

[1] Romer, "The Welcoming of Copernicus's *De revolutionibus*," 181.
[2] Translations from the *Commentariolus* are adapted from Rosen, *Three Copernican Treatises*, 135, in consultation with Swerdlow, "The derivation and first draft of Copernicus' Planetary Theory," and with the original Latin text in Prowe, *Nicolaus Coppernicus*.

The preamble to the *Commentariolus* takes up only a page, after which appears Copernicus's list of seven "postulates" or "axioms"—a list that presents the Copernican Universe in a nutshell.

Axiom 1

"There is no one center of all the celestial circles or spheres."

Though it sounds startling, this axiom is actually the least radical of the seven. The eccentrics of Ptolemy already conceded that different circles had different centers. Copernicus never doubted, however, that the heavenly motions must be circular. Kepler with his elliptical orbits would later show otherwise.

Axiom 2

"The center of the Earth is not the center of the Universe, but only of gravity and of the lunar sphere."

Ptolemaic astronomy with its eccentrics virtually agreed with the first part of this Copernican axiom. But with the mention of gravity (which in a sixteenth-century context just means "heaviness," the thing that makes heavy stuff like rocks go down), whole new possibilities open up, possibilities that came to fruition a century and a half after Copernicus in the work of Isaac Newton.

Axiom 3

"All the spheres revolve about the Sun as their midpoint, and therefore the Sun is near the center of the Universe."

This astonishing claim, the signature assertion of the new cosmology, is slightly more vague than it sounds in some modern translations, for Copernicus too retained a version of eccentrics. Thus, in his view, the Sun is merely very near (*circa*) the center of the Universe, with the spheres of Mercury, Venus, Earth, Mars, Jupiter, and Saturn revolving about it.

Axiom 4

"The distance from Earth to the Sun is imperceptible in comparison with the height of the firmament."[3]

Among the seven axioms, this may be the most surprising in its consequences, even if on casual first reading it sounds merely abstract. According

---

[3] A précis of the axiom. Copernicus's words here: "The ratio of Earth's distance from the Sun to the height of the firmament is so much smaller than the ratio of Earth's radius to its distance from the Sun that the distance from Earth to the Sun is imperceptible in comparison with the height of the firmament."

to Ptolemy, the Universe—that is, the sphere of the fixed stars that we see surrounding Earth when we look up at night, sometimes referred to as the "firmament"—is so large "that Earth has the ratio of a point to the heavens."[4] Ptolemy supported this conclusion *empirically* by adducing the fact that a plane drawn through an observer's line of sight at any point on Earth always bisects the whole heavenly sphere. In other words, if we study the stars, we find that we always are seeing half of the sphere of stars, at least if we have a clear horizon. But we should see less than half because Earth's surface, our standpoint for observing, is off-center by the distance between the center of Earth and its surface (i.e., by one Earth radius). Ptolemy therefore concluded that the Universe is *immensely* (literally, *immeasurably*) vaster than the Earth; and conversely, in relation to the Universe, Earth is immensely small, as nothing (see Figure 2.2).

However, Copernicus knew that in his system an observer's standpoint on Earth's orbit around the Sun *likewise* appears to bisect the sphere of the fixed stars. According to Axiom 4, the height of the firmament—the distance from Earth to the sphere of the fixed stars—is therefore *so* immense that the Earth–Sun distance itself is also as nothing by comparison. It was well known that the Earth–Sun distance is very large indeed—far larger

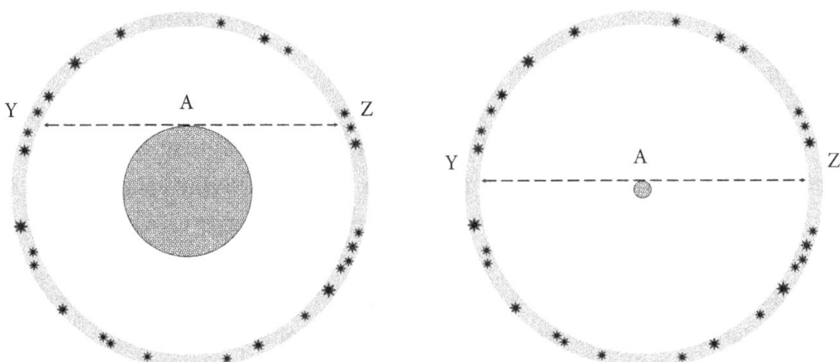

Figure 2.2 Left: Under Ptolemy, an astronomer located on Earth at A can only see the stars above the horizon (dashed line), between Y and Z. Right: If Earth is small, then the astronomer sees nearly half the sphere of stars. If Earth is so small as to be but a point, then the astronomer will see exactly half.

---

[4] Ptolemy, *Almagest*; adapted from *BOTC* [*The Book of the Cosmos: Imagining the Universe from Heraclitus to Hawking*, ed. Danielson], 72.

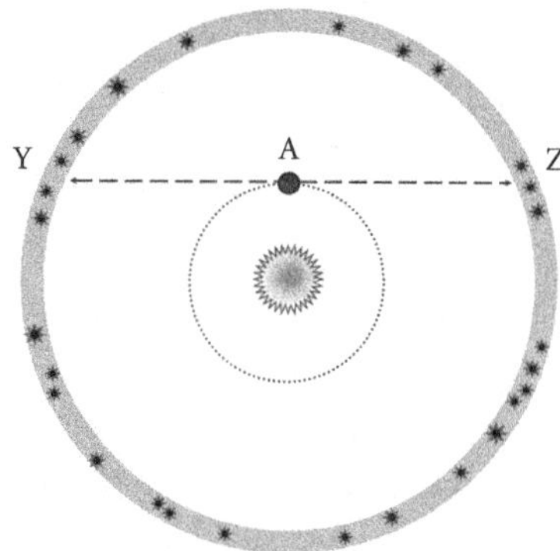

Figure 2.3 Under Copernicus, the size of the orbit of the Earth about the Sun (dotted circle) replaces the size of the Earth itself in the "half the sphere of stars" question.

than the Earth. How enormous, then, must be that immensity of the starry sphere which makes "very large" appear as nothing! In short, Copernicus's Axiom 4 established a radical new benchmark for the immensity of the Universe (see Figure 2.3).

This most surprising conclusion—perhaps as much as his preposterous claims regarding Earth's motions—would, as we'll see, create huge obstacles to the acceptance of Copernicus's system for a century or more.

Axiom 5

"Whatever motion appears in the firmament arises not from any motion of the firmament, but from Earth's motion. Earth together with its accompanying elements performs a complete rotation on its fixed poles in a daily motion, while the firmament and highest heaven abide unchanged."

This axiom, marking what is usually called the rotation of the Earth, dealt a coup de grâce to one of the hardest-to-imagine tenets of Ptolemaic astronomy: the claim that the entire immense stellar sphere makes a full revolution every day, carrying all the celestial machinery of epicycles and eccentrics with it, causing the daily rising and setting of Sun, Moon, and everything else in the sky.

Axiom 6

"What appears to us as motion of the Sun [through the constellations] arises not from its motions but from the motion of Earth and our sphere,

with which we revolve about the Sun like any other planet. Earth has, then, more than one motion."

One of the most familiar Copernican assertions, this axiom opened up a whole new way of "reading" the book of Nature, according to which the state and motion of the perceiving subject (i.e., the earthly observer) is really what accounts for the apparent state or motion of the object of perception. In this case, the object is the Sun, which apparently moves through the Zodiac constellations over the course of a year, but this apparent motion is simply a reflection of Earth's yearly revolution or orbit about it.

Axiom 7

"The apparent retrograde and direct motion of the planets arises not from their motion but from the Earth's. The motion of Earth alone, therefore, suffices to explain so many apparent inequalities in the heavens."

Continuing the scientific explanation of "observer–subjectivity" from Axioms 5 and 6, this axiom offered to solve at a stroke one of the great embarrassments of Ptolemaic astronomy: that, despite its strenuous efforts, it could not describe the heavens in a way that rescued their regularity and uniformity of motion. Under Copernicus, there *were* no irregularities, no "inequalities." There were only *apparent* ones. Mars made no *real* reversals in its motion; by Axiom 3, its real motion was to circle the Sun, just like Earth. Mars only made *apparent* reversals, caused by Earth's periodically passing between it and the Sun. Copernicus could therefore explain what Ptolemy could not: why the retrograde motion of Mars always occurred when it was opposite the Sun in the sky.

So, with Axioms 3 and 7 together, Copernicus was stating, for the first time in human history, that Mars (along with Venus, Mercury, etc.) had something in common with Earth. All revolved around the Sun. None actually cycled backward in their motions. *Earth was a planet.* And if Earth were a planet ("like any other planet," said Axiom 6), then so also might the planets be other earths. This was the germ of an idea that would quickly lead to us Earthlings imagining that, like Earth, other planets might be inhabited with intelligent life. There might be Martians, Jovians, and other extraterrestrials.

Regardless of how radical this list of axioms was, Copernicus ended here by reminding his readers that he was attempting to frame a system that would achieve what astronomers had been seeking all along: an absence of real "inequalities" in the heavens. What he longed for instead was a comprehensive, harmonious vision of a truly unified Universe.

One of Copernicus's most enduring contributions was his persistent assumption that real science must not only be consistent with evidence but also be elegant and beautiful: It had to be "pleasing to the mind." For him, this insistence is at one with the assertion that "the heavens declare the glory of God" of Psalm 19:1—which meant that an aesthetically inferior cosmology would not reflect well on the Creator. In his letter to the Pope at the head of his published treatise, *Revolutions* (*De revolutionibus*, 1543), Copernicus complained that his predecessors, by inserting all those epicycles and eccentrics, "have been like someone attempting a portrait by assembling hands, feet, head, and other parts from different sources. These several bits may be well painted, but they do not fit together to make up a single body. Bearing no genuine relationship to each other, such components, joined together, would compose a monster, not a man.... It began to irritate me that the philosophers [that is, scientists] . . . could not agree on a more reliable theory concerning the motions of the system of the Universe, which the best and most orderly Artist of all framed for our sake."[5]

And part of the beauty of Copernicus's Cosmos is the elegance of the whole, as indicated by his famous diagram from Book I of the *Revolutions* (see Figure 2.4). Notably, the period of each planet increases in accordance with its distance from the Sun: Mercury, eighty-eight days; Venus, nine months; Earth, one year; Mars, two years; Jupiter, twelve years; Saturn, thirty years. And the whole celestial machinery, including the fixed stars, no longer has to spin about daily! It is now *immobile*. This arrangement, Copernicus suggests, is beautiful, "pleasing to the mind." It offers a harmoniously unified Universe. And in the words of his student Rheticus, it is "not unworthy of God's workmanship."[6]

At another level, Copernicus's radical revision of the cosmic model can be seen as almost modest. Here, side by side, are schematic depictions of the old Universe and the new Universe (Figure 2.4). On the left is the two-storey Universe. At its center, enclosed by the sphere of the moon, is our world. This is an Earth-centered, or *geocentric*, Universe. Our world is composed of the elements mentioned by Copernicus in Axiom 5 that in his time were thought to comprise it: earth and water in the middle, air (and stylized clouds) above them, encircled by a sphere of fire (again, with stylized flames). That's the lower storey, "downstairs." Then comes the perfect

[5] *BOTC*, 106.
[6] Rheticus, *Narratio Prima*; Rosen, *Three Copernican Treatises*, 145.

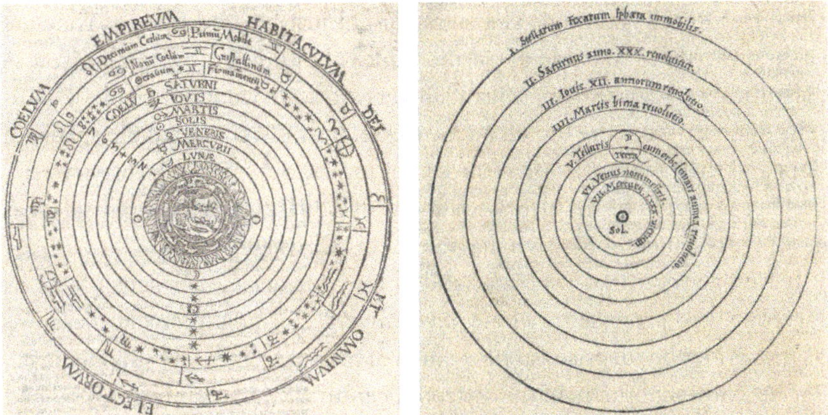

Figure 2.4 A diagram of the Pre-Copernican Universe (from Peter Apian's sixteenth-century *Cosmographicus*) and Copernicus's famous diagram. Note that Copernicus's diagram does not show the distance from Earth (*V. Terra*) to the sphere of the fixed stars (*I. Stellarum Fixarum Sphæra immobilis*) being so immense that the Earth–Sun distance itself is as nothing by comparison, contrary to his fourth axiom.
Credit: ETH-Bibliothek Zürich.

and unchanging "upstairs," the planetary spheres of Moon, Mercury, Venus, Sun, Mars, Jupiter, Saturn (Uranus and Neptune not yet discovered), all surrounded by the eighth sphere, that of the fixed stars or "firmament," followed by the ninth, "crystalline" sphere, superimposed by the signs of the Zodiac, in turn encompassed within the "Primum Mobile" that impels the spheres below it. Surrounding all is the "empyreal" Heaven, the "dwelling of God and of all the elect."

But if we stop at the starry sphere and shift our gaze right to the diagram of the new Cosmos, the main innovation appears disarmingly simple: Copernicus has merely swapped around the positions of Earth and Sun. We do notice that the starry sphere is now labeled "immobile"—*stellarum fixarum sphaera immobilis*. The revision does not appear radical: The circles are still circles; the planets are still embedded in their "spheres" (not following "orbits," as they would in the following century); and the starry sphere is still bounded, finite.

Now take a second look at that starry sphere in Copernicus's diagram. Do you see a problem? In Ptolemy's Universe, the starry sphere was indeed located as diagrammed: just beyond Saturn. Not so in Copernicus's

Universe! Recall Axiom 4. The stars are not just beyond Saturn. They are so far away that the circle of Earth's orbit is but a point by comparison. A diagram showing the Copernican Universe to scale would in fact look radically different from the Ptolemaic Universe. It would consist of the sphere of stars and then merely a dot at the center for the Sun and planetary system. Perhaps it should be no surprise that one of Copernicus's most radical ideas is not represented in his own diagram of his Universe.

If we overlook that problem, and if we consider Copernicus's fame as someone who proposed a whole new Universe, *we* might look at these two diagrams and be surprised at how much of his model is in fact still medieval. For his contemporaries in the sixteenth century, however, his proposal had enough novelty to make minds spin. The Universe no longer has upper and lower storeys. Earth is now a wandering star—a planet—one that in addition rotates on its axis daily (how does it do that?). *And* it revolves about the Sun annually, with the Moon in its own orb tagging along with mothership Earth independently from the other planets. And the Sun: Deprived of its celestial status—no longer a planet!—it is now to be located in the center, the cosmic basement. As expressed in the words of Giovanni Maria Tolosani, who published the first critique of the *Revolutions*, Copernicus "puts the indestructible Sun in a place subject to destruction."[7]

Among historians of astronomy, it has become something of a parlor game to try to list everyone before the year 1600 who might truly have accepted Copernicus's cosmology. Nobody gets beyond ten or twelve names. Initially, to accept the Copernican model as a description of how things actually are in the Universe was akin to believing six impossible things before breakfast. There *were* those who managed it. But it took a lot of imagination, much hard work, and decades of vigorous debate and careful examination of evidence before many more could join Copernicus in taking that giant leap.

---

[7] Tolosani, writing in June 1544, "Tolosani's Condemnations of Copernicus' *Revolutions*," 189.

3

"Little Dark Star"

In 1540, three years before publication of the great *Revolutions*, Georg Joachim Rheticus, a young mathematician from the University of Wittenberg and Copernicus's only student, wrote a sort of trial balloon summarizing Copernicus's cosmology. Hoping to convince his teacher that there was an audience for his ideas, Rheticus then sent copies of the book, titled simply *A First Account*, to influential people across central and northern Europe. One of them, Rheticus's former teacher Achilles Gasser of Lindau, in southwest Germany, in turn forwarded his copy to Georg Vögeli, a friend living on the other side of Lake Constance. Gasser included an introductory letter offering a vivid glimpse of how wonderful—and how difficult to credit—the new cosmology would inevitably appear to its early readership. The letter begins: "Behold, most distinguished sir, I am sending you . . . this little book. Not only is it something new and unknown to our contemporaries, but also (if I am not mistaken) it will surprise and utterly astonish you even to the verge of disbelief."[1]

Gasser's letter and the whole text of *A First Account* were republished in 1566 as an appendix to the second edition of Copernicus's *Revolutions*. One copy of this volume then found its way to England and into the hands of another brilliant mathematician, whose name was Thomas Digges (1546–1595). On the volume's title page Digges boldly wrote: *Vulgi opinio Error*—"the common opinion is wrong"—and went on to become the foremost English Copernican of his century. In 1576, at the age of only thirty, he published the first ever translation of Copernicus from Latin into a vernacular language—yes, into English—only eighteen years into the reign of Elizabeth I and twenty years before Johannes Kepler publicly declared for Copernicanism.

Four years earlier, in 1572, Thomas Digges had been one of a handful of astronomers to observe a stunning celestial phenomenon that perhaps more than any other single event undermined the two-storey cosmology that

---

[1] Danielson, "Achilles Gasser and the Birth of Copernicanism," 460.

A Universe of Earths. Dennis Danielson and Christopher M. Graney, Oxford University Press.
© Oxford University Press (2025). DOI: 10.1093/9780197803547.003.0003

Copernicus's model eventually replaced. On November 11 of that year, 26-year-old Danish astronomer Tycho Brahe (1546–1601) was observing the constellation Cassiopeia and noticed a new star. So thorough and perceptive was the Dane's account of the phenomenon—which made *him* as well as the star famous—that still today it is known as Tycho's Supernova (albeit scientists in their poetic way prefer to call it SN1572).

It was brilliant. It passed high overhead as seen from Tycho's Denmark. And it was far brighter even than anything in the sky except the Moon and Sun, bright enough indeed to see during daylight. To say Tycho was surprised at the new star's appearance would be an understatement. "I was so astonished at this sight," he wrote, "that I was not ashamed to doubt the trustworthiness of my own eyes"[2]—eyes that would become the best trained and most trustworthy of any astronomer in all Europe.

Tycho was amazed not merely to see a new star but also then to grasp what it meant for the nature of the Universe. He went on to offer a scientific account of his observations over the course of the following year, the most striking of which was that the star retained its position among the other fixed stars. Even the Moon is close enough to Earth that its location seen against the stars of the Zodiac is slightly altered by the place from which it is observed (see Figure 3.1). This phenomenon is called *parallax*. The new star showed no parallax, meaning it was much more distant than the Moon. It represented change, stunning, brilliant change, taking place in the upper storey of the Cosmos, "in the ethereal region of the celestial world," as Tycho wrote, where for almost two millennia Aristotle and his followers had declared that "*no* change … takes place."[3] This new star guaranteed that the days of the two-storey cosmology were numbered.

Thomas Digges, too, observed the new star of 1572, and the following year he declared, also like Tycho, that the new phenomenon is "far beyond the sphere of the moon."[4] Those observations together with Digges's discovery of Copernicus clearly felt to him like an epiphany—and yet the young Englishman knew that conveying an epiphany to others might require huge amounts of diplomacy and delicate persuasion. People don't just willingly or instantly change their minds about long, deeply held views of the world.

---

[2] See Brahe, *De Nova Stella*, 13 (*BOTC* [*The Book of the Cosmos: Imagining the Universe from Heraclitus to Hawking*, ed. Danielson], 129).

[3] See Brahe, *De Nova Stella*, 13 (*BOTC*, 129). Italics added.

[4] Digges, *Alae seu scalae mathematicae*, sig. A.iii.ᵛ.

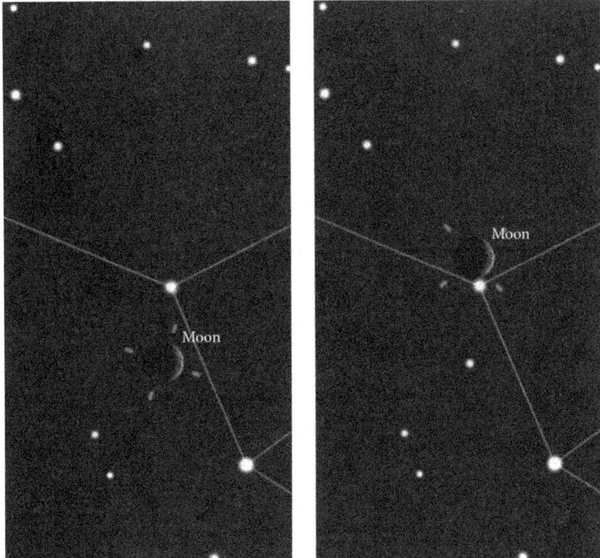

Figure 3.1 The Moon among the stars of Sagittarius, as seen from southern Chile (left) and from Hudson Bay in Canada (right) on November 6, 2024. The apparent change in the position of the Moon that results from the change in observing location is called *parallax*.
Images made using the *Stellarium* computer application.

They need something to hold them back from what Gasser had called "the verge of disbelief."

Beginning in the 1550s, Thomas Digges's father, Leonard Digges, had published a pamphlet called *A Prognostication Everlasting*, a sort of perennial almanac with rules about how to judge the weather using the Sun, Moon, stars, and other celestial phenomena. But by 1576 it was time for Thomas to correct some of the errors that had crept into his father's work, and he noticed that the *Prognostication* contained a schematic drawing of "the situation of the spheres celestial and elementary according to the doctrine of Ptolemy, whereunto all universities (led thereunto chiefly by the authority of Aristotle) since have consented." Thomas decided to add a Copernican drawing to *Prognostication*.

In this work Thomas Digges doesn't outright dismiss Aristotle. On the contrary, he leaves his father's schematic of the Ptolemaic Universe in place and intact. But in his glowing introduction of Copernicus's astonishing new cosmology, Thomas holds nothing back:

But in this our age, one rare wit (seeing the continual errors that from time to time more and more have been discovered, besides the infinite absurdities in their theories, which they have been forced to admit that would not confess any mobility in the ball of the Earth) hath by long study, painful practice, and rare invention delivered a new theory or model of the world, showing that the Earth resteth not in the center of the whole world, but only in the center of this our mortal world or globe of elements, which environed and enclosed in the Moon's orb, and together with the whole globe of mortality is carried yearly round about the Sun, which like a king in the midst of all reigneth and giveth laws of motion to the rest, spherically dispersing his glorious beams of light through all this sacred celestial temple. And the Earth itself to be one of the planets, having his peculiar and strange courses turning every 24 hours round upon his own center: whereby the Sun and great globe of fixed stars seem to sway about and turn, albeit indeed they remain fixed. So many ways is the sense of mortal men abused.

But reason and deep discourse of wit having opened these things to Copernicus, and the same being with demonstrations mathematical most apparently by him to the world delivered: I thought it convenient together with the old theory also to publish this, to the end such noble English minds (as delight to reach above the baser sort of men) might not be altogether defrauded of so noble a part of philosophy. And to the end it might manifestly appear, that Copernicus meant not (as some have fondly excused him) to deliver these grounds of the Earth's mobility only as mathematical principles, feigned and not as philosophical truly averred.[5]

It's a stunning piece of writing. Digges flatters his "noble" English audience, permits them to keep thinking of Earth as located at the center of the sublunary sphere—of "this our mortal world or globe of elements." Indeed, elsewhere he speaks of how "in the midst of this Globe of Mortality hangeth this dark star or ball of earth and water."[6] At the same time he gives them an arresting taste of Copernican vertigo with Earth following "peculiar and strange courses" while the stars "seem to sway about and turn, albeit indeed they remain fixed." On the one hand Digges makes huge concessions to traditional Aristotelian cosmology, elements and all (earth, water, air, fire), while

---

[5] Digges, "To the Reader," *A perfit description of the Caelestiall Orbes*, sig. M.1.[r–v].
[6] Digges, *A perfit description*, sig. M.2.[r]. See also N.4.[r]: "this little dark star wherein we live."

on the other hand stating flat out that Copernicus's system is no fiction. Digges knows that many ("fondly" means foolishly) have regarded Copernicus as a producer of excellent mathematics, while treating his cosmology as a mere useful hypothesis. But no, that's not right: Copernicus's system is "philosophical truly averred"—a delightful pleonasm that essentially means "truly truly true." In effect, Thomas says this is how the Universe actually functions.

Yet the more "conservative" aspect of Thomas Digges's Universe appears most clearly if we examine his father Leonard's graphic of the Ptolemaic world picture together with the new Copernican one, as we did in the last chapter (see Figure 3.2). The former is familiar—the sublunary four elements: earth and water, above which float air, in the form of stylized clouds, and, above this, fire, with stylized flames reaching up to the sphere of the Moon.

But now shifting our gaze to Thomas Digges's diagram of the new Copernican Universe, we are struck by its remarkable hybridity, hybridity that may in fact have helped keep sixteenth-century readers' minds open to the astonishing Universe here being introduced. Copernicanism was hard to accept because it required that so very many ingrained assumptions be set aside. This impediment is famously summed up by John Donne's memorable lament, uttered thirty-five years later, in 1611, that the

> new philosophy calls *all* in doubt,
> The element of fire is quite put out.

However, Digges, as a translator and champion of Copernicus, deftly *avoids* putting out the element of fire, or eliminating any of the other elements essential to Aristotelian sublunary physics, which to Digges's audience would have seemed little more than common sense.

In Book 1, Chapter 10 of *Revolutions*, Copernicus wrote that "**this whole domain encircled by the Moon,** with the center of the Earth, traverses this great circle [*orbis magnus*] amid the other wandering stars in an annual revolution around the Sun." Digges freely translates the bolded phrase of this sentence as "**this whole globe of elements enclosed with the Moon's sphere,**"[7] even though Copernicus himself in this context says nothing at all about elements. Digges retains not only such Aristotelian sublunary

---

[7] Digges, *A perfit description*, sig. N.2.ᵛ.

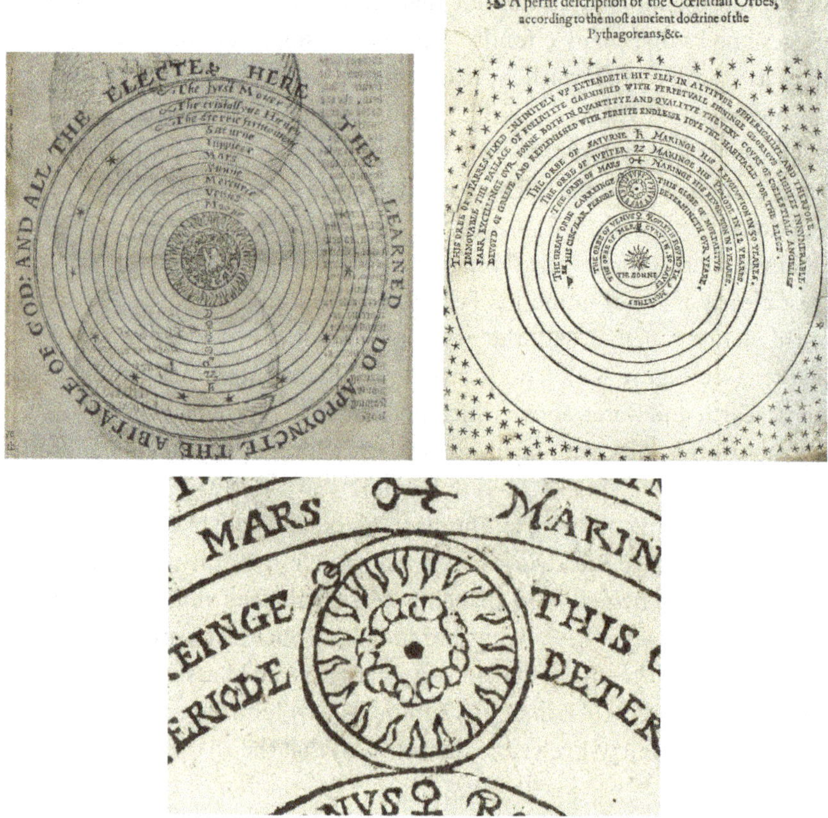

Figure 3.2 The Ptolemaic (left) and Copernican (right) Universes as illustrated by Thomas Digges. Again, note that the Copernican diagram does not show the distance to the stars correctly; they should be so far away as to make the Earth's orbit look like a point. Digges graphically transposed (bottom) the sublunar realm of earth, water, air, and fire from Aristotle and Ptolemy onto the Copernican system.

Credit: The Linda Hall Library of Science, Engineering & Technology and Wikimedia Commons/Wellcome.

physics, but also the traditional implication that this "elementary" domain is uniquely a "globe of mortality." In effect, Digges simply lifts that elemental "globe of mortality" from his father's Ptolemaic picture and transposes it—earth, air, fire, and all—into its new location on the great circle of the Copernican system, photoshops the familiar sublunary world onto the recently enlarged picture of the *super*lunary cosmos offered by Copernicus.

Thus, there, orbiting the Sun, is the sphere of the Moon, the stylized clouds, the flames reaching up to the lunar sphere (see Figure 3.2, bottom). And thus his readers can feel right at home in the new Universe without feeling they have to leave their old familiar cosmic neighborhood.

What Digges's picture is best known for, however, is an innovation on a much larger scale: all those stars extending themselves out spherically beyond the edge of the page, along with his superinscribed phrase declaring "This orb of stars fixed infinitely up." Yet like the Aristotelian elements, the diffused and infinite domain of the stars—stars that Digges describes as "far excelling our Sun both in quantity and quality"—is nowhere to be found in Copernicus.

We don't know Digges's precise source for this radically different conception of the starry realm. It was only later, in 1584, that Giordano Bruno would publish his dialogue *On the Infinite Universe and Worlds*. But in 1563 there appeared in Paris an edition of the ancient atomist poet Lucretius's *On the Nature of Things*, whose first book declares the Universe to be infinite. So it's not inconceivable that Digges's infinite starry sphere might derive directly or indirectly from that direction. In any case, the infinite world of ancient Atomism became an increasingly well-known thesis in the second half of the sixteenth century and beyond—and it truly opened the door wide to speculation about a "Plurality of Worlds" and their possible inhabitants, as we shall see.

In 1614 Johann Locher, a student of the Jesuit astronomer Christoph Scheiner, published *Mathematical Disquisitions Concerning Astronomical Controversies and Novelties*, a survey of astronomical discoveries and ideas that included the infinite universe diagram-plus-description seen here[8] (see Figure 3.3). Locher labels this the "system of the ancients" and discusses it, but he makes clear that he doesn't accept or promote it. Writing that "truth and Christian philosophy have long since rejected such fictions in favor of something that has stability,"[9] Locher uses the diagram as a starting point for mathematical arguments *against* the possibility of any physical infinity, arguments that prefigure the twentieth-century "Hilbert's Hotel" arguments of the mathematician David Hilbert.

Of the three circles that appear in Locher's graphic, the outer one is of course merely a pictorial convenience fitted to a finite page. The "INFINITE

---

[8] Locher and Scheiner, *Disquisitiones Mathematicae*, in *MathDisq* [Locher with Scheiner, *Disquisitiones Mathematicae*, tr. Graney], 17.

[9] "Disquisition 9," *MathDisq*, 17.

Figure 3.3 Locher and Scheiner's diagram of an infinite Universe from *Disquisitiones Mathematicae*, 1614.
Credit: ETH-Bibliothek Zürich.

CHAOS OF ATOMS" is by definition *not* bounded. The next circle (ABCD), bounding the "Starry Heavens," indicates the separation of a finite sidereal sphere from the surrounding (theoretically) uncircumscribed Chaos. Finally, in this schematic, the innermost circle merely demarcates the starry realm from the "place of the planets and of Earth" (note that Scheiner and Locher mention Earth separately). So the inner circle is left completely blank, or rather black. Potentially any arrangement of the planets and Earth—geocentric or heliocentric—could simply be cut-and-pasted into this bounded but otherwise undefined realm.

The picture illustrates how, within forty years of Digges's Copernican publication, significant changes in the modeling of the super-planetary realm were being recognized, if often not accepted. We notice in Locher's picture not only the deepening of the starry sphere (much thicker than in Copernicus's schematic) but also the connection of a cosmic infinitude with the resurgent doctrines of Atomism. The endpoint of such changes would be the complete abandonment of disputes about what occupied the center of the Universe. In Lucretius's words:

> For since the void is infinite, the space
> Immense, how can there be a middle place?[10]

It's not the last time we'll hear questions of this sort.

But for Thomas Digges, back in 1576, there *was* a middle place, and it was occupied by the Sun, which—echoing the language of Copernicus himself—"like a king in the midst of all reigneth and giveth laws of motion to the rest, spherically dispersing his glorious beams of light through all this sacred celestial temple." And the Earth, with the whole "globe of mortality," is to be reimagined as a planet, a wandering star, circling it, yet a "*dark* star"! And at the other spatial limit of Digges's fast and loose eclecticism, as we've seen, is that picture-worth-a-thousand-words of all those stars extending themselves out spherically beyond the edge of the page, "fixed infinitely up . . . excelling our Sun both in quantity and quality."

Digges's achievement is no less a Copernican landmark despite his offering a modular, hybrid, cosmic mashup: an amalgam of models that permits his audience to cling to a familiar, premodern sublunary world—to an Earth now planetary and stellar but still dark—which then gets dropped into a harmonious new Copernican Universe, which itself is then quietly but dramatically inserted into an infinite super-Cosmos.

Digges might be right that the "common opinion is wrong," but he doesn't discard all of it. And he offers much else with which that opinion—and thoughtful sixteenth-century English minds and imaginations—can be challenged, surprised, and breathtakingly augmented.

---

[10] Lucretius, *De Rerum Natura*, Bk. I, 33; *BOTC*, 29.

4

Galileo

"Dance of the Stars"

It will come as no surprise that the first ever astronomical observations made with a telescope produced a series of breathtakingly "Aha!" moments. Galileo Galilei (1564–1642) made six such landmark observations beginning in late 1609: (1) the rugged, Earth-like surface of the Moon; (2) the additional countless multitude of stars that now came into view; (3) the stellar constitution of the Milky Way and of certain nebulae (small, luminous clouds), whose component stars could now be resolved; (4) the moons of Jupiter; (5) the phases of Venus; and (6) the spots that appeared on the face of the Sun. Galileo documented the first four of these observations in a little book called *Starry Messenger*, which he rushed into print in the first half of 1610. Beginning this book with a brief description of how he heard about and successfully reverse-engineered the telescope, Galileo then launches straight into a more systematic account of how he used his device to observe first the Moon and "afterwards . . . both fixed stars and planets, with incredible delight."

Galileo's contagious excitement at making his discoveries—even if some historians of astronomy question whether in each case Galileo was the first to make them—is doubled by his quick grasp and eloquent accounts of what they *mean*. With his telescope, he can see spots (Latin *maculae*) on the surface of the Moon, and he can distinguish the "great or ancient" ones visible to the naked eye from other, smaller spots that "sprinkle the whole surface of the Moon" and are seen only through the telescope. But right away, Galileo both claims scientific precedence and draws a dramatic inference regarding Earth's place in the Universe:

These spots have never been observed by anyone before me; and from my repeated observations I have arrived at the following conclusion: that we undoubtedly do not perceive the surface of the Moon to be perfectly smooth, free from inequalities and exactly spherical (as a large school

A Universe of Earths. Dennis Danielson and Christopher M. Graney, Oxford University Press.
© Oxford University Press (2025). DOI: 10.1093/9780197803547.003.0004

of philosophers believes concerning both the Moon and other heavenly
bodies), but on the contrary to be full of inequalities, uneven, full of hol-
lows and protuberances. It is like the surface of the Earth itself, which is
everywhere varied with lofty mountains and deep valleys.[1]

So: no more two-storey Universe! No more perfection upstairs and imper-
fection downstairs. The Moon is Earth-like; its surface is not *immaculate*!
Galileo knows immediately that his telescopic observations are delivering a
whole new picture of the Cosmos—and perhaps of Earth within it.

The Moon was not the only celestial body that the telescope revealed
to be less than immaculate. The telescope showed that the Sun also was
spotted. Galileo, with some justification, receives credit for discovering this
solar phenomenon, but Christoph Scheiner deserves recognition for his
observational and interpretive contribution as well.

In 1611, himself having taken up a telescope and begun observing *maculae* on the face of the Sun, Scheiner wrote three letters on the subject, offering
a preliminary account of his findings. He sent these to his friend Marc Welser
in Augsburg, who published them and sent them to Galileo for his response.
From the beginning of the letters, Scheiner speaks respectfully of the Sun
as "the source of illumination and the commander of heavenly bodies." But
only a few sentences farther along Scheiner identifies the specific problem:
As he was using the "optical tube" to measure magnitudes of the Moon and
the Sun, he "noticed on the Sun some rather blackish spots like dark specks."[2]

As with telescopic observations of the Moon, the discovery of "spots" had
a more-than-scientific resonance. "Spot" could imply blemish, a moral or
metaphysical fault—as even today we might speak of something being a
"spot" on someone's reputation. If this sense of the word caused a problem
for Galileo's account of the surface of the Moon, it seemed, for Scheiner,
even more disturbing when applied to the Sun. For this reason—because it is
"unfitting . . . that on the most lucid body of the Sun there would be spots"—
Scheiner interprets the appearances not as something *on* the solar surface
but instead as stars (i.e., planets) circling about in the Sun's "sky," analogous
to what Galileo had discovered orbiting Jupiter. In his third letter, Scheiner

---

[1] Galileo, *Siderius nuncius* (Venice, 1610), fol. 7ᵛ. All translations of this work are adapted from
The Sidereal Messenger of Galileo Galilei, translated by E. S. Carlos (London, 1880); excerpted in
BOTC [*The Book of the Cosmos: Imagining the Universe from Heraclitus to Hawking*, ed. Danielson],
145–54.

[2] Galilei and Scheiner, *On Sunspots*, 59, 61.

returns to the theme of championing the Sun's honor by means of a planetary interpretation of the spots: "It pleases me to liberate the Sun's body entirely from the insult of spots."[3] Nor was he alone in expressing shock—as much moral and symbolic as cosmological—at the possibility that the glorious Sun's face might not be immaculate. In a jocular but perceptive letter dated January 4, 1612, the Tuscan poet Alessandro Allegri acknowledged that the "Flemish glass" (the telescope) was being used to reveal "filth on the cheeks of the Sun."[4]

For his part, Galileo resolutely—and, from a scientific point of view, admirably—refused to let this rhetorical or symbolic dimension of the discussion interfere with his treatment of plain evidence. In his response to Welser, publisher of Scheiner's *Three Letters*, he carefully rejects a priori assumptions concerning where sunspots might actually be located:

> The author [of *Three Letters*], having established that the observed spots are not illusions of the lens or defects of the eye, seeks to determine something general about their location, showing that they are neither in the air nor on the solar body. . . . But the hypothesis that they cannot be on the solar body does not appear to me to have been fully and necessarily determined. For it is not conclusive to say, as he does in the first argument, that because the solar body is very bright it is not credible that there are dark spots on it, because as long as no cloud or impurity whatsoever has been seen on it we have to designate it as most pure and most bright, but when it reveals itself to be partly impure and spotted, why shouldn't we call it both spotted and impure?[5]

The antipathy Scheiner expressed regarding the possibility of a besmirched Sun is instructive, but he was actively revising his position in the face of new data. By 1614, when he and Locher wrote their *Mathematical Disquisitions*, Scheiner would say that

> *The spots are blackish bodies, roving around the Sun by various motions. Thus far neither their number nor their nature has been determined.* They are so close to the sun that they seem to be attached to it. . . . Whether the spots are planets is currently disputed and will be disputed for some time.

[3] Galilei and Scheiner, *On Sunspots*, 62, 67.
[4] Galilei and Scheiner, *On Sunspots*, 77; Allegri, *Lettere di Ser Poi Pedante*, 21.
[5] Galilei and Scheiner, *On Sunspots*, 91.

Scheiner and Locher recommended the writings of Galileo and said to "Expect more in due time."[6]

Scheiner did produce more in due time. Galileo commends his openness to evidence: The author of *Three Letters* "begins to lend his ear and approval to the true and good philosophy, and especially in the matter concerning the constitution of the universe, but . . . he cannot yet totally free himself from those fancies previously impressed on him, . . . habituated to assent by long custom."[7] In 1630 Scheiner published a monumental study of the Sun, based on systematic observations made over years. He described sunspots and their nature and movement in detail. Scheiner was the astronomer who would hone the use of a telescope for in-depth, systematic astronomical research, and his object of study was the Sun. Still, he was right to say in 1614 that the nature of the spots would be disputed for some time: Otto von Guericke's 1672 diagram of the Solar System showed the spots as bodies orbiting the Sun (see Figure 1.1, right—where the diagram shows sunspots [*maculae*, c] as bodies orbiting the Sun inside the orbit of Mercury). So half a century after the work of Galileo, even a Copernican might still cling to the idea of a spotless Sun in a world that had already moved beyond a two-storey Universe.

Spots on the Sun and Moon were not the only telescopic discoveries that undermined the two-storey Universe with its "down here" and "up there." Galileo's observations of lunar mountains did as well, all the more so thanks to his brilliant feat of calculating lunar elevations. Samuel Edgerton describes Galileo's calculation of the height of a mountain on the Moon's dark side whose top peeps up out of the Moon's shadow to be bathed in sunlight:

Since the moon's diameter was known to be two-sevenths of the earth's diameter, or about 2,000 miles, Galileo . . . revealed by Pythagorean calculation that . . . the mountain's height on center from its base reached more than four miles into the lunar sky! By applying a problem well known to students of Renaissance perspective, Galileo added yet another fact to his already wondrous revelations, that the mountains on the moon were more spectacular than the Alps here on earth.[8]

[6] "Disquisition 30," in *MathDisq* [Locher with Scheiner, *Disquisitiones Mathematicae*, tr. Graney], 76–77.

[7] Galilei and Scheiner, *On Sunspots*, 95.

[8] Edgerton, *The Heritage of Giotto's Geometry*, 246.

The calculation is remarkable not only for Galileo's accuracy given the limitations of his equipment, but also for the very fact that he was able to apply *geometry*—literally "earth measure"—extraterrestrially. The implication is that not only the landscape but also space itself is qualitatively the same "up there" as it is "down here." Edgerton calls this the "geometrization of astronomical space."

The extraterrestrial application of geometry also, of course, dated from at least Ptolemy's argument that Earth "has the ratio of a point to the heavens" thanks to the plane of the horizon bisecting the sphere of the stars. More recently, it had played a role in Digges's and Tycho's efforts to measure parallax in the supernova of 1572. Copernicus's student Rheticus conceived of the astronomical applicability of geometry as symbolic (though not *merely* symbolic) of the unity of creation, just as Copernicus conceived of the Universe as possessing a fundamental symmetry (a word that again implies a common measure). Thus Rheticus—subtly echoing the Lord's Prayer (*sicut in cælo, et in terra*)—points to "God's geometry in heaven and on Earth."[9] In the same way, Galileo knew his successful use of geometry as a universal yardstick lent significant support to a vision of a single homogeneous Cosmos created by a sole geometer God. Little wonder that Kepler, upon hearing of Galileo's discoveries, would exclaim that "geometry . . . shines in the mind of God."[10]

But to Galileo, "the connection and resemblance between the Moon and the Earth" extends beyond mountains and geometry. Earth too shines, he says. It illuminates the Moon.

When the Moon is a crescent, he says, we do not see a crescent only. We see the rest of the Moon, too, glowing with a faint light—"a slight and faint circumference is also seen to mark out the circle of the dark part, that part, namely, which is turned away from the Sun." We do not need a telescope to see this, but if a telescope is used, Galileo says, even those spots are visible on the "dark side."

What is the source of the "dark side" light, the "secondary light" of the Moon? It is the Earth, Galileo says. When the Moon is seen as a crescent from Earth, the Earth would be seen as nearly full from the Moon. And as the full Moon lights up Earth at night, so the full Earth lights up the Moon. Thus not only is the surface of the Moon geometrical and Earth-like, but

---

[9] Burmeister, *Georg Joachim Rhetikus, 1514–1574*, 3:139.
[10] *KCGSM* [Kepler, *Kepler's Conversation with Galileo's Sidereal Messenger*, tr. Rosen], 43.

there is also a reciprocal relationship between Moon and Earth as shining astronomical companions.

This is a radical attack on the two-storey Universe. When Locher and Scheiner reviewed Galileo's argument in 1614, they noted that Gerolamo Cardano in his 1554 book *On Subtlety* had said that, in the upper storey, "all things shine and glitter to such a degree, that if indeed we might inspect the Moon, being there during the time of an eclipse, we might be blind on account of a splendor not unlike innumerable brilliantly lit candles focused into the eyes."[11] In an eclipse!—when the Sun is blocked from illuminating the Moon! In other words, the Moon, like everything else in the upper storey, is inherently brilliant. The Sun just makes it more so. Were Galileo right, Locher and Scheiner said, then a person on the Moon could read by the light of the Earth, whereas on Earth no one can read by the light of the stars—meaning the Earth would outshine all the stars together![12]

The importance of this idea of Earth outshining the stars can hardly be overstated for an Earth previously thought to be qualitatively set apart—virtually quarantined within the "sublunary sphere"—indeed, considered no astronomical body at all. We recall Digges, who *did* consider Earth an astronomical body, yet called it a "dark star." Galileo, however, envisaged the Earth shining in the lunar sky, illuminated by sunlight, just as we see the sunlit Moon shining in our sky.

This "Earthshine" is an interesting enough phenomenon on its own, but its cosmological implications are stunning. It implies a Cosmos in which light travels a two-way street. For Galileo, it evinces a kind of friendly commerce that also has immense consequences for humanity's conception of the place and role of Earth in the Universe at large:

The Earth, with fair and grateful exchange, pays back to the Moon an illumination like that which it receives from the Moon nearly the whole time during the darkest gloom of night. . . . . The benefit of [the Moon's] light to the Earth is balanced and repaid by the benefit of the light of the Earth to her. For while the Moon approaches the Sun about the time of the new Moon, she has in front of her the entire surface of that hemisphere of the Earth which is exposed to the Sun and vividly illumined with his beams, and so receives light reflected from the Earth. Because of this reflection, the

---

[11] "Disquisition 27," in *MathDisq*, 69.
[12] "Disquisition 27," *MathDisq*.

hemisphere of the Moon nearer to us, though deprived of sunlight, appears of considerable brightness.... This is the law observed between these two orbs: Whenever the Earth is most brightly enlightened by the Moon, that's when the Moon is least enlightened by the Earth, and vice versa.

Galileo promises that in a subsequent work he will offer further evidence

to demonstrate a very strong reflection of the Sun's light from the Earth—this for the benefit of those who assert, principally on the grounds that it has neither motion nor light, that the Earth must be excluded from the dance of the stars. For I will prove that the Earth does have motion, that it surpasses the Moon in brightness, and that it is not the sump where the Universe's filth and ephemera collect.[13]

As we'll see later, this last paragraph is crucially important given the dubious contention that ancient and medieval geocentrism placed Earth and humankind in a position of supreme importance in the Universe. Galileo's comment indicates exactly the contrary: that the center, as a place where heavy, gross things settle, was seen as a place of disrepute, and a place distinctly different from the rest of the Universe. Therefore, Copernicanism decisively entails Earth's *exaltation*, pulling it out of the filth, making it like the rest of the planets, a shining wandering star—decidedly not a sump or *sentina*, the term Galileo uses in his original Latin to evoke the place in a ship below the lowest decks, where bilge water collects. For him, Earth belongs not there but in the "dance of the stars."[14]

Galileo's telescope said something about the fixed stars, too: They are innumerable. From the time of ancient Babylon to early seventeenth-century Europe, there had been various estimates regarding the number of the fixed stars, which were generally held to be uncountable (for practical reasons), but in reality were finite in number, perhaps a few thousand. Keen eyes saw more stars, weak eyes saw fewer.[15] Yes, Digges and various atomists had believed them to be infinite, but until the telescope there was absolutely no *empirical* support for any such claim—and of course there still wasn't.

---

[13] Fol. 15r–16r: *non autem sordium, mundanarumque fecum sentinam*; BOTC, 150.

[14] According to historian of astronomy Noel Swerdlow, "None of Galileo's discoveries provoked more hostility and more preposterous attempts at refutation than the rough surface of the Moon and the explanation of the secondary light, and with good reason because for none were the stakes as high"; Machamer, *Cambridge Companion to Galileo*, 250–51.

[15] As late as 1640, John Wilkins suggests the remarkably low number of 1022 stars. *Discourse Concerning a New Planet*, 56.

But to go from two or three thousand to *innumerable* was a major shift in our scientifically attested view of the starry realm.

With a mixture of awe and triumph, Galileo continues to tell his audience both what they will see when they too use the telescope and what consequences their discovery will have for "disputes that have tormented philosophers for so many centuries." Point the telescope, he says, at the great white path in the night sky—the Milky Way, the "Galaxy" (from the Greek for milk; think of *lac*tose): "The Galaxy is nothing else but a mass of innumerable stars planted together in clusters. Point your telescope in any direction within the Galaxy and at once a great mass of stars comes into view. Of these, many are fairly large and robustly conspicuous, but the number of small ones is utterly unfathomable."[16] Galileo's Milky Way was not a luminescent pathway, but a pattern of points and spaces, or grains and voids. The Universe thus appeared through Galileo's telescope more and more as an array of discrete bits—as suggested by Donne's complaint that the new science left the world "all in pieces, all coherence gone."[17]

Galileo gestured triumphantly toward Earth's inclusion and participation in "the dance of the stars"—stars that were, as the telescope showed, innumerable. Against the whole pre-Copernican background, this was something for Earth's inhabitants to savor, something cosmologically dynamic, elevating, exhilarating. And we Earthlings may savor it still.

---

[16] *BOTC*, 152.

[17] *An Anatomy of the World: The First Anniversary*, line 213; Donne, *The Complete English Poems*, 276.

5

"Nest of Angels"

Early Modern Extraterrestrials

The landmark developments in mathematical and observational astronomy in the century following the death of Copernicus in 1543 drove speculation about life, possibly intelligent life, elsewhere in the Universe. Tycho's, Digges's, and Galileo's achievements in eliminating the divisions between what had been considered the Universe's upper and lower storeys made it possible to think that "up there" might be something like "down here." The speculation might be summed up by the words of Milton's angel Raphael in 1667:

> what if Earth
> Be but the shadow of heaven, and things therein
> Each to other like, more than on Earth is thought?
>
> (*Paradise Lost* 5.574–76)

Roughly: "What if heaven and Earth are more alike than you on Earth imagine?" In short, the erasure of the boundary between the lower and upper storeys of the Universe implied that these domains no longer needed to be thought of as radically distinct, or characterized by utterly dissimilar physics, substances, and beings.

We have already noticed the hybridity of Digges's contribution, with Earth remaining as a "little dark star"[1] within a Copernican model that is in turn nested in an infinite Universe of stars. There's a similar hybridity in a poet whom Digges quotes repeatedly in the pages preceding his translation of Copernicus, Marcellus Palingenius Stellatus (1500–1551?). This poet's *Zodiacus vitae*, although in many ways a philosophical hodgepodge, must have appealed to sixteenth-century English Protestant readers. It was printed in England seven times in Latin between 1569 and 1599 and three

---

[1] *A perfit description*, sig. N.4.[r].

A Universe of Earths. Dennis Danielson and Christopher M. Graney, Oxford University Press.
© Oxford University Press (2025). DOI: 10.1093/9780197803547.003.0005

times in Barnaby Googe's English translation (1565, 1576, 1588). One wonders if its reputation was also perhaps enhanced by its appearance in the first Catholic *Index of Prohibited Books*. For our purposes, it is notable not only for perpetuating the familiar gloomy estimation of life in the sublunary sphere ("all that nature framed beneath the Moon, is nought, and ill"), but also for repeatedly proposing the likely existence of extraterrestrials. Palingenius argues this claim from the immense largeness of the Universe relative to the Earth ("the seas and earth … are, compared to the skies, as nothing"), as well as from the quasi-religious belief in the "Principle of Plenitude": that God's creativity tends to fill all the places God creates. Of the Earth, accordingly, Palingenius writes:

> Shall then so small and vile a place so many fish contain
> Such store of men, of beasts and fowls and th'other void remain?
> Shall skies and air their dwellers lack? He dotes that thinketh so.

He concludes therefore that "Each star an island shall be thought," and "doubtless heaven, stars, and air inhabitants enjoys."[2] Palingenius's logic of air dwellers perhaps implies creatures more like angels than like us, but nevertheless it is uncannily resonant with the oft-repeated maxim of Carl Sagan and other proponents of SETI, that "If it's just us [in our enormous Universe, then it] seems like an awful waste of space."[3]

Palingenius was neither an astronomer nor a Copernican, yet his writings are evidence that talk of extraterrestrial life in an immense Universe had already appeared prominently on the scene in the sixteenth century some decades before the writings of Giordano Bruno, whom general accounts often give disproportionate credit for introducing such ideas.[4] Digges, whose work also preceded both Bruno's and Kepler's, was indeed an astronomer and a Copernican. Yet, as already indicated, he deftly fused Copernicanism with a familiar vocabulary inherited from the Aristotelian/Ptolemaic model, repeatedly referring to Earth as a "dark star."[5]

---

[2] Palingenius, *The Zodiake of Life*, sig. GG.vii.^{r}, sig. X.v.^{r}, sig. X.vi.^{r}

[3] Most dramatically stated in a discussion between the characters played by Jodi Foster and Matthew McConaughey while overlooking the Arecibo radio telescope in the 1997 movie *Contact*, based on Carl Sagan's book of the same name. See our further discussion of this scene in Chapter 16.

[4] On, for example, Bruno's tenuous grasp of Copernicanism, see McMullin, "Bruno and Copernicus."

[5] Digges, *A perfit description*, sig. M.2^r.

As far as extraterrestrial life is concerned, the crucial word here is simply "star." We have seen how Copernicanism did not, as many have assumed, *demote* Earth. In the two-storey Universe, Earth had sat in the sump of the Universe, or, in the choice words of Giovanni Pico (1463–1494), the "excrementary and filthy parts of the lower world."[6] Copernicanism *demoted the Sun* from its planetary status and *exalted the Earth* into the heavens.[7] Even today, scientists as well as the general public often only dimly grasp the extent to which Copernicanism thus raised, not lowered, the cosmic status of Earth; and it did so in part by theorizing Earth as a light-bearer and light-sharer (along with the Moon and other planets). It did so, in other words, by making Earth part of a dynamic cosmic community—no longer, to repeat Galileo's words, "excluded from the dance of the stars." In the old cosmology, the planets were identified with divinity, as their names still today suggest—Venus, Mars, Jupiter, etc.—something to which the upstart Copernican Earth now cheekily seemed to aspire. Most decisively for our discussion here, the re-conception of Earth as a planet, a wandering *star*, established a firm and fruitful analogy between it and the rest of the Universe. This cosmological "homogenization"—the establishment of a strong physical uniformity between the Universe "up there" and the earthly realm "down here"—meant that in a Copernican Universe, the Earth/stars analogy was a two-way street: Not only was Earth re-conceived as a planet, but also the other planets could now readily be imagined as other earths, homes not to air dwellers but to beings like us.

The Englishman John Wilkins (1614–1672) in his 1638 *The Discovery of a World in the Moone* would emphasize the homogeneity or uniformity of space. Within this space, he declared, the twin planets of the Earth and Moon inhabit "but one region" and enjoy a mutually beneficial light-sharing relationship "as loving friends."[8] The Aristotelian/Ptolemaic dark, isolated, sump-like Earth could hardly have undergone a more radical imaginative transformation, nor one of greater consequence for humankind's capacity to imagine—and perhaps form relationships with—beings elsewhere in the Universe.

Johannes Kepler instantly recognized these implications. Upon reading Galileo's *Starry Messenger* in 1610, he immediately asserted the probability "that there are inhabitants not only on the moon but on Jupiter too." As we have seen, he went on to speculate that the "Jovians" may enjoy four moons

---

[6] Pico, "Oration on the Dignity of Man," 224. The original phrase is "excrementarias ac foeculentas inferioris mundi partes" (Pico, *Opera Omnia*, 314).

[7] More on this theme in Chapter 7.

[8] Wilkins, *The Discovery of a World in the Moone*, 153.

(unlike us, who have only one) as consolation for the fact that they are less ideally located in the Universe than we Earthlings are. Kepler also boldly prophesied the day when we might launch our own lunar and planetary expeditions. For surely "settlers from our species of man will not be lacking"; and "given ships or sails adapted to the breezes of heaven, there will be those who will not shrink from even that vast expanse."[9] Similarly Robert Burton, in his encyclopedic *Anatomy of Melancholy* in 1621, recognizing the demise of Aristotelian physics and the abolition of ethereal spheres, endorsed the *sine qua non* of space travel: that "If the heavens be penetrable . . . it were not amiss in this aerial progress to make wings, and fly up."[10]

Moreover, the achievements of Copernican astronomers such as Galileo and Kepler led to what historian David Cressy has called "England's lunar moment."[11] In 1638, two influential works appeared that helped awaken more thoughts of space travel than ever before. The first of these, published posthumously, was Francis Godwin's imaginative fiction *The Man in the Moon: or a Discourse of a Voyage Thither*. Some elements of Godwin's narrative, such as the tethered flock of geese that conveys the main character to the Moon, are fanciful. But the journey offers a vivid, non-Aristotelian account of physical features such as gravitation as well as the daily rotation of the Earth "according to the late opinion of Copernicus." What Godwin's fiction perhaps most movingly conveys, however—something actualized powerfully and photographically in the late 1960s by the Apollo missions—is a vision of our own planet as a "new star" masked "with a kind of brightness like another moon."[12]

The second was John Wilkins's book, *The Discovery of a World in the Moone*, where he extrapolates from his enthusiastic presentation of Copernican astronomy the idea of an inhabited Moon. Like Godwin, Wilkins not only vividly describes conditions on the Moon but also imagines the shining appearance of our native globe from space. With some relish, he supports the scientific credibility of this imagined vision by citing the authority of two contemporary anti-Copernican Continental philosophers:

> Thus also Carolus Malapertius, whose words are these . . .
> "If we were placed in the Moon, and from thence beheld this
> our Earth, it would appear unto us very bright, like one of the

---

[9] *KCGSM* [Kepler, *Kepler's Conversation with Galileo's Sidereal Messenger*, tr. Rosen], 39. Kepler extended his brilliant exploration of possible lunar travel, environment, and perspectives much further in his posthumous *Somnium: The Dream*.

[10] Burton, *The Anatomy of Melancholy*, 325. See also Danielson, "Ancestors of Apollo."

[11] Cressy, "Early Modern Space Travel and the English Man in the Moon," 967.

[12] Godwin, *The Man in the Moon*, 90, 92.

> nobler planets." Unto these doth Fromondus assent, when he
> says . . . "I believe that this globe of Earth and water would
> appear like some great star to anyone who should look upon
> it from the Moon."[13]

Not all anti-Copernicans granted this. Locher and Scheiner in their 1614 *Disquisitions* have a whole section and large diagram on "Different moons." They note that the moons of Jupiter seen from there would look as our Moon does to us here on Earth, progressing from crescent to full and back. But whereas our Moon does this in a month, the four Jovian moons that Galileo discovered would do so in varying periods of time: the innermost moon in a day and a half, they write; the outermost moon in seventeen days. Indeed, they assert, were a man to be placed on Saturn, he might observe "all the planets to be images of moons"—but they do not mention Earth as a planet, and do not show an Earth in their diagram.[14]

Wilkins admits the difficulties of a journey to the Moon; yet, like others, he builds on the recent success of journeys to earthly places such as America (another kind of "New World"). And thus, he concludes by eloquently echoing the prophetic strains of Kepler regarding ships to the Moon. He admits he can't conjecture how one might sail to the Moon. "We have not now any Drake or Columbus to undertake this voyage, or any Daedalus to invent a conveyance through the air. However, I doubt not but that time who is still the father of new truths . . . will also manifest to our posterity that which we now desire but cannot know."[15]

In Milton's *Paradise Lost* (1667)—which John Tanner has called "perhaps the greatest description of space travel in high-brow fiction"[16]—we hear hints of

> every star perhaps a world
> Of destined habitation (7.621–622)

and observe the anti-hero Satan as an astronaut (literally a sailor among the stars). In the second-to-last stop on his journey to tempt humankind,

---

[13] Wilkins, *The Discovery of a World in the Moone*, 149–50.

[14] "Disquisition 43," in *MathDisq* [Locher with Scheiner, *Disquisitiones Mathematicae*, tr. Graney], 102–103.

[15] Wilkins, *The Discovery of a World in the Moone*, 107. The other prominent mid-seventeenth account of a lunar voyage was that of Cyrano de Bergerac, *L'Autre Monde*, posthumously published in 1657.

[16] Tanner, "'And Every Star Perhaps a World of Destined Habitation': Milton and Moonmen," 268.

the Adversary alights on the Sun to ask directions of its resident angel. The angel offers him a thoroughly Galilean prospect of Earth, which appears as a globe that "shines" and so *can be seen*, just like the other wandering stars, in particular like its

> neighboring Moon
> (So call that opposite fair star) (3.727–28).

In the earlier words of Robert Burton, Earth "shines to them in the Moon, and to the other planetary inhabitants, as the Moon and they do to us."[17]

Repeatedly in the seventeenth century, therefore, both scientists and poets not only looked outward into a newly conceived Universe but also exercised "reflexive telescopics," a phrase coined by Hans Blumenberg in the 1970s—shorthand for the imagined examination of Earth from an extraterrestrial viewpoint, the natural complement of terrestrial observation of, and speculation concerning, what is "out there." As Blumenberg writes, no sooner had Galileo trained his telescope upon the Moon than the question arose, How would Earth appear through a telescope?[18] At the end of this chapter, we'll return to one of the most detailed and beautiful instances of such a scenario, one written by Thomas Traherne in the 1670s: that of a dweller among the stars approaching our Earth and remarking on what must be the condition of its inhabitants.

But first let's practice some reflexive telescopics ourselves, beginning with a brief interpretation of the frontispiece that appeared in John Wilkins's 1640 *Discourse Concerning a New World & Another Planet*, a re-publication in a single volume of two works he had published separately. The *New World* discourse is simply his 1638 *The Discovery of a World in the Moone*, an extrapolation from the works of both Galileo and Kepler regarding the nature of the Moon as an earthlike planet with, by analogy, earthlike inhabitants. Again, every analogy has two sides, and the "discovery" of an earthlike Moon was thus appropriately followed in the second book by an examination of what Wilkins called "another planet"—by which of course he meant planet Earth, still apparently a novel-sounding idea even a century after the publication of Copernicus's *Revolutions* (see Figure 5.1).

As for Copernicus, here he is in Wilkins's frontispiece posing the hypothesis that in real life he *never* posed *merely* as a hypothesis. He borrows Tycho Brahe's trademark question: *Quid si sic*, "What if it be thus?" But on

---

[17] Burton, *Anatomy of Melancholy*, 326–27.
[18] Blumenberg, *The Genesis of the Copernican World*, 675.

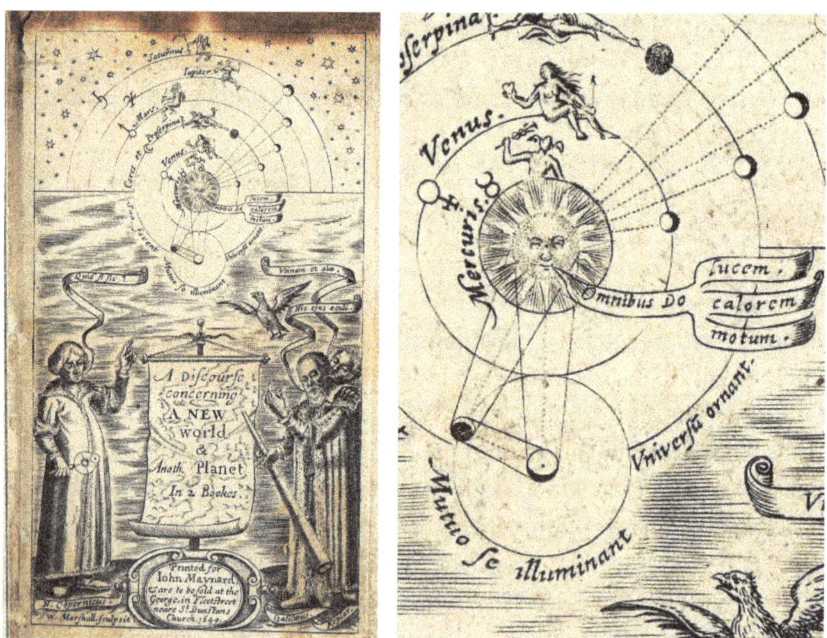

Figure 5.1   The Wilkins frontispiece (1640, left). As in the Copernican diagrams seen previously, no attempt is made here to convey the vast distance to the Copernican stars. The diagram of the Sun and planets in the frontispiece shows the reciprocal luminosity between Earth and Moon (right). Credit: Archive.org.

the opposite side of the title stand Wilkins's other two heroes, Galileo and Kepler, who represent the twin pillars of astronomy then as now: 1. Observation (Galileo with his telescope, "Here be his eyes"); and 2. Mathematics (Kepler: "Yes, and his wings"). The other thing not to be missed in the top third of Wilkins's frontispiece is that the Copernican Cosmos offered here is the *English* Copernican Cosmos inherited from Digges, with the stars not in a nice neat band or stellar sphere, but spilling off over the edge of the page and so suggesting a possibly infinite Universe.

Nonetheless, even amid that immensity there is a coherent family of Sun-and-planets which in Wilkins's charming but serious cartoon can be seen shining on each other (see Figure 5.1, right). The Sun "gives light, warmth, and motion" (*lucem, calorem, motum*) to them all, while the Moon and Earth, in keeping with their reciprocal luminosity as advocated by Galileo, "enlighten each other" (*Mutuo se illuminant*). No wonder that Kepler, Wilkins, and their inheritors, having postulated beings on the Moon and having arrived at the basic but decisive Galilean realization that Earth is visible from "out there," began imagining how Earth might appear to Moon dwellers.

Giordano Bruno, the Copernican so famous for his embrace of an infinite Universe of "innumerable suns, and an infinite number of earths revolv[ing] around those suns," also considered how things appear both to those dwelling on the Moon ("which is another earth") and to those dwelling elsewhere. "Those who dwell on the moon and on the other stars," he wrote in his 1584 *On the Infinite Universe and Worlds*, would all see their world as "the very center of the universe . . . immobile and fixed, the whole universe revolv[ing] around her."[19] Thus, all the Universe's inhabitants are star-dwellers, seeing themselves as fixed at the Universe's center, yet capable of imagining how things appear to those on the innumerable other stars.

This idea of a Universe of innumerable stars orbited by innumerable inhabited earths, all part of the dance of the stars, would seem to be the logical end to the process of cosmological homogenization that we have discussed so far. It is a potent idea. It undergirds many aspects of modern popular and scientific culture. The starship *Enterprise* boldly going to encounter new life and new civilizations on numerous "Class M" planets; a seedy cantina, frequented by a dizzying array of intelligent life forms, where Obi-wan hires Han Solo to provide transport in his *Millennium Falcon*; the similarly dizzying array of life in the Marvel Cinematic Universe—all these blockbuster entertainment franchises, now familiar to people on every continent of Earth, presume a Universe full of other earths. So does the world of UFO (or, to use the more recent term, UAP) enthusiasts and "space alien" conspiracy theories—which can impinge on serious scientific work, as illustrated by NASA's UAP panel in 2023, which felt the need to acknowledge questions like "what is NASA . . . and the science overlords . . . hiding and where are you hiding it?"; and indeed which opened discussion by addressing the problem of the harassment of panelists by those who are sure that NASA is hiding evidence of extraterrestrials.[20] So does the world of science itself, as seen in SETI and in serious scientists proposing that objects passing through the Solar System (e.g., ʻOumuamua, the unusually-shaped object that passed through in 2017) might be the products of an extraterrestrial civilization.

---

[19] Crowe, *The Extraterrestrial Life Debate: Antiquity to 1915*, 46.
[20] "Public Meeting on Unidentified Anomalous Phenomena," at 3:24:20 and 00:03:45.

But not all Copernicans were of the opinion of Bruno. We saw in Chapter 1 how Kepler wrote of "our little ball, the little cottage of us all, which we call the Earth . . . the architect of marvelous works," inhabited by "these fine bits of dust called human beings . . . in whom is the image of God; who are, in a certain way, lords of the whole bulk." Kepler seems willing only to take the cosmological homogenization so far. He thinks Earth is *not* just another world. And in a manuscript discovered only in the 1990s but dating from the 1670s, Thomas Traherne offers the scenario of a "Celestial Stranger," an example of ETI, discovering our noble planet for the first time. Traherne's brilliant imagination clearly sees Earth as a "star" within the Universe, and certainly not the Universe's sump, but also not just another world:

> Had a man been always in one of the stars, or confined to the body of the flaming Sun, or surrounded with nothing but pure ether, at vast and prodigious distances from the Earth, acquainted with nothing but the azure sky and face of heaven, little could he dream of any treasures hidden in that azure veil afar off. . . . Should he be let down on a sudden, and see the sea, and the effects of those influences he never dreamed of; such strange kind of creatures; such mysteries and varieties; . . . such never heard of colors; such a new and lively green in the meadows; such odoriferous and fragrant flowers; such reviving and refreshing winds, . . . it would make him cry out How blessed are thy holy people, how divine, how highly exalted! Heaven itself is under their feet! . . . The Earth seems to swell with pride, that it bears them all; all its treasure[s] laugh and sing to serve them. . . . Verily this star is a nest of angels! . . . This little star so wide and so full of mysteries! So capacious, and so full of territories, containing innumerable repositories of delight, when we draw near! Who would have expected, who could have hoped for such enjoyments?[21]

It's a powerful question that includes, but goes far beyond, powerful sentiments. And it's a question that resonates still.

---

[21] Traherne, *Poetry and Prose,* 112–14.

6

Star Amid the Darkness

A book about the discovery of Earth as a planet requires some attention to background, both background of the very notion of *planet* and literal visual background. We all know that what "sets off" the stars as we scan a clear night sky is the blackness that surrounds them. So, did you notice that blackness does not feature in Thomas Traherne's discussion (at the end of the previous chapter) of what a man at "vast and prodigious distances from the Earth" would see? Not black, but azure: "the azure sky"; "that azure veil."

William Shatner, who played *Star Trek*'s Captain Kirk, saw both blue and black when he went into space in 2021, at the age of 90, on Jeff Bezos's rocket *New Shepard*. Upon returning to Earth, he spoke of this: "The covering of blue. This sheet, this blanket, this comforter that we have around. We think, 'Oh, that's blue sky,' . . . Then suddenly you shoot through it all of the sudden, as though you're whipping a sheet off you when you're asleep, and you're looking into blackness, into black ugliness."[1]

Shatner speaks of breaking through the blue to find himself in the blackness. Traherne imagines no blackness. Not just the Earth, but sky and space have been radically reconceived across the centuries. That reconception affects our understanding of this our home planet, as Shatner's remarks illustrate.

Copernicus made our home planet a "star." How do we think about stars? Even the colloquial use of the word "star" to describe someone who's outstanding in his or her field—whether it be sports star, movie star, or rising star in any given profession—suggests that we place a superlative evaluation on someone or something we call a star. Astronomically, today we normally use the word to denote a self-luminous celestial body. As we've already noticed, from ancient times to the early modern period what we now call stars were referred to as *fixed* stars. This distinguished them in all their regularity and predictability from the *wandering* stars or *planets*, those other

---

[1] Hernandez, "William Shatner boldly went into space."

A Universe of Earths. Dennis Danielson and Christopher M. Graney, Oxford University Press.
© Oxford University Press (2025). DOI: 10.1093/9780197803547.003.0006

bright lights that moved against the heavenly background of the fixed stars of the Zodiac.

Now we know that planets are *not* self-luminous. It would therefore be tempting for us to treat these wandering stars either not as stars at all or else as stars of lesser status, pretenders to stardom, as it were. Ancient and other observers right up to early modern times faced no such temptation. Then as now, when it comes to sheer brightness in our sky, Jupiter and Venus easily outshine Sirius, the brightest of the fixed stars; Mars can too, at times. So to the naked eye at least, there's no obvious reason for considering the wandering stars to be lesser or dimmer objects than the fixed ones. This claim is emphatically true for two other heavenly bodies we now seldom think of as planets: the Sun (which *is* self-luminous) and the Moon. Both of these bodies—like the other planets—appear to move about in the sky against the background of the fixed stars of the Zodiac. These, classically, make up the seven planets, each still honored in various languages as they feature in the names of the seven days of the week:

- Sunday—Sun
- Monday—Moon
- Martes—*Spanish*, Mars
- Mercredi—*French*, Mercury
- Jovedì—*Italian*, Jove (Jupiter)
- Venerdì—*Italian*, Venus
- Saturday—Saturn

These seven "planets" (Uranus and Neptune are not visible to the naked eye and were discovered telescopically) were not only honored in many languages but also, across ages and cultures, often objects of veneration, even worship. They were considered literally heavenly beings, divine. A notable exception to this pattern appears in the Hebrew Scriptures. The well-known account in Genesis of the fourth day of creation simply avoids specific names for the Sun and the Moon, referring to them only as "lights" created "to give light upon the earth," a "greater light" and a "lesser light" (1:14–16). Scholars have suggested that biblical writers avoided the Hebrew noun for the Sun—*shemesh*—because it "was too reminiscent of the Canaanite and Mesopotamian Sun god, Shamash."[2] In Judaism, idolatry—worship of "other

---

[2] Miller, "'Epigraphical' Rabbis, Helios, and Psalm 19," 61.

gods"—was to be avoided at all costs. Similarly in the book of Job, the title character "judges it necessary to profess that he never raised his hand in homage to the sun or the moon" (Job 31: 26–27), although he too avoids the word *shemesh*, employing instead the same word for "light" used in the first chapter of the Bible.[3]

Even in religious and cultural contexts in which the Sun was not considered a divinity, it still seemed to warrant the honor of being considered heavenly. This, as we noticed in Chapter 2, posed a problem for Copernicus, whose model removes the Sun from what was considered the heavens and locates it in the lowest point in the Cosmos, "a place," the early critic Tolosani complained, "subject to destruction."[4] Copernicus's response was to devote some of his most poetic language to establishing that the Sun's central place in his model was *not* a place of dishonor.

> Behold, in the midst of all resides the Sun. For who, in this beautiful temple, would set this lamp in another or a better place, whence to illuminate all things at once? For aptly indeed do some call him the lantern—and others the mind or the ruler—of the Universe. Trismegistus calls him the visible god, and Sophocles' Electra 'the beholder' of all things. Truly indeed does the Sun, as if seated upon a royal throne, govern his family of stars as they circle about him.[5]

To extend a point made in Chapter 2, one to which we'll return, Copernicus thus takes pains, rhetorically, to renovate what had been considered the cosmic basement in order to present it as a fit dwelling place for the Sun.

The counterpart, of course, to the Sun's transition to the Universe's lowest point was Earth's rise to the heavenly realm, its promotion to planetary status—*star* status. If that seems counter to what you have heard about Copernicus, bear with us; we'll soon return to what we call the great Copernican cliché, the oft-repeated myth that Copernicus's cosmology "demoted" Earth from its supposedly privileged position at the cosmic center. But suffice for now to note that the new model, by declaring Earth to be a planet, a wandering *star*, repositioned it into the heavens. It's true, as we've seen, that an early Copernican such as Digges, for all his audacity, continued to

---

[3] Lipínski, "Shemesh," 764–68.

[4] Tolosani, "Tolosani's Condemnations of Copernicus' *Revolutions*," 189.

[5] Copernicus, *De revolutionibus*, I.x. Adapted from the translation that appears in *BOTC* [*The Book of the Cosmos: Imagining the Universe from Heraclitus to Hawking*, ed. Danielson], 117.

see this now exalted planet as a "little *dark* star." But the next generation of Copernicans recognized that the newly dynamic planetary Earth "surpasses the Moon in brightness" (Galileo) and "shines to them in the Moon, and to the other planetary inhabitants, as the Moon and they do to us" (Burton).[6]

Perhaps no single picture more graphically illustrates Earth's exquisite Copernican brightness than NASA's "Blue Marble" image, one of the most widely reproduced photos in history (see Figure 6.1). This is no "dark star." Its glory is not diminished by our knowing that its light derives from Copernicus's Sun located amid its "beautiful temple" there at the cosmic center.

Figure 6.1 The "Blue Marble."
Credit: NASA

[6] *Galileo in BOTC*, 150; Burton, *Anatomy of Melancholy*, 326–27.

This image is cropped in Figure 6.1 to foreground Earth against the inky blackness of space. That blackness is very much part of the picture, and, without the severe cropping, it, not Earth, would overwhelmingly dominate the picture. The comments of NASA astronauts, especially members of the lunar missions, who viewed our home planet from afar, illustrate this disproportion: notably, Gene Cernan (Apollo 10 and 17): "You're looking across the blackness of space a quarter of a million miles away, looking at the most beautiful star in the heavens . . . and it's moving in a blackness that's almost beyond conception." Neil Armstrong (Apollo 11), trying to catch some sleep at Tranquility Base on the lunar surface, "was kept awake by the Earth shining . . . 'like a big blue eyeball'; [yet] it was so small that he could blot it out with his thumb." Michael Collins (Apollo 11) indicated not just the depth of the darkness but also its near omnipresence as he and his crew, beginning their homeward journey from the Moon, tried visually to spot Earth. "The little planet is so small out there in the vastness that at first I couldn't even locate it. And then I did. . . . There it was, shining like a jewel in a black sky."[7]

But Copernicus, whom we honor for declaring Earth's planetary status, for making it a star, could not have imagined seeing planet Earth like a jewel in a black sky for the simple reason that he had no conception of a thoroughly black sky. The intellectual historian Vladimir Brljak has recently undertaken a sweeping study of, as he puts it, "a shift in the European cosmological imagination whose significance is potentially comparable to those from geocentrism to heliocentrism, or a bounded to an unbounded universe, yet whose cultural impact remains almost entirely uncharted."[8] There does not seem to be a specific moment at which this shift took place. Essentially, however, before the shift, the sky and indeed the whole Universe beyond Earth were thought to be bright and blue, illuminated by the Sun—Traherne's azure Universe. After the shift, the Universe was recognized as dark and black. And accordingly, before the shift, night was considered to be a temporary darkening of the brightness and blueness; after the shift, it is the brightness of our day that is the cosmological anomaly.

---

[7] Astronauts' comments quoted from Poole, *Earthrise*, 99–100.

[8] Vladimir Brljak has generously shared with us significant amounts of his as-yet unpublished work (to be published with Reaktion Books as *When Did Space Turn Dark?* See https://www.durham.ac.uk/staff/vladimir-brljak). Many of the resources we cite derive from Brljak's research. These ideas have also been discussed by Clifford Cunningham, especially in his June 21, 2019 presentation on "When Did Man First Realise That Space Is Black?" at the Biennial History of Astronomy Workshops University of Notre Dame XIV (https://www3.nd.edu/~histast/workshops/2019ndxiv/abstracts.shtml), specifically discussing Figure 6.3.

The before-the-shift understanding of night is best illustrated by some pictures. The thirteenth-century image presented in Figure 6.2 shows that the shadow created by the Earth accounts for the darkness of our night. That shadow obscures the bright blueness of what we have come to call space. According to this conception, says C. S. Lewis in *The Discarded Image*,

Night is merely the conical shadow cast by our Earth. It extends, according to Dante (*Paradiso*, IX, 118) as far as to the sphere of Venus. Since the Sun moves and the Earth is stationary, we must picture this long, black finger perpetually revolving like the hand of a clock; that is why Milton calls it "the circling canopie of Night's extended shade" (*Paradise Lost*, III, 556). Beyond that there is no night; only "happie climes that lie where day never shuts his eye" (*Comus*, 978). When we look up at the night sky we are looking through darkness but not at darkness.[9]

Figure 6.2 The shadow of Earth as a black zone in a blue Universe (left). Getty Museum, Ms. Ludwig XV 4 (83.MR.174), fol. 150. 13th century. At right are two illustrations showing the phases of the Moon (fols. 149, 148). Note that the superlunary region is blue but the sublunary region is not; also note the brilliant gold of the Sun and Moon versus the dull gray of Earth. https://www.getty.edu/art/collection/object/108E3S.

[9] Lewis, *The Discarded Image*, 111–12.

The fifteenth-century image shown in Figure 6.3 portrays the planets in shining gold, against, once again, a *blue* background. Earth is not golden (and, interestingly enough, the Moon is darker still). The lovely manuscript seen in Figure 6.4 likewise shows the Universe as blue.

Yet, contrary to Lewis's eloquent account, which seems to tie the shift in our view of darkness to the shift from geocentrism to heliocentrism, heliocentrism did not automatically force a change in that view. Copernicus himself explicitly held to the earlier conception of night, writing that "while the rest of the universe is bright and full of daylight, night is clearly nothing but the earth's shadow, which extends in the shape of a cone and ends in a point."[10]

Brljak offers this concise summary of the earlier view: "We know that if we take off in a spaceship at noon, we will emerge into eternal night.

Figure 6.3 The blue of the Universe in Christianus Prolianus, 1478, manuscript "Astronomia" (Latin MS 53) at Manchester University, https://luna.manchester.ac.uk/luna/servlet/detail/Man4MedievalVC~4~4~836375~136541?page=119&qvq=&mi=119&trs=162.

[10] Copernicus, *On the Revolutions*, 177.

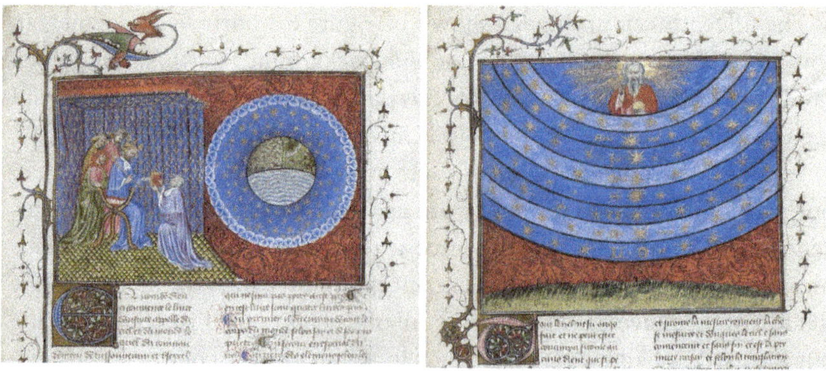

Figure 6.4 Representations of the Universe as blue from the manuscript "Nicole Oresme, Traité de la sphère; Aristote, De caelo et de mundo, traduction française par Nicole Oresme," at Bibliothèque nationale de France, Département des Manuscrits (Français 565, c. 1410), https://gallica.bnf.fr/ark:/12148/btv1b8451098g/f55.planchecontact

They thought that if they took off at midnight, they would emerge into eternal day." Or note another of Brljak's summaries of the contrast: For us, "daylight is the anomaly, and darkness the norm," whereas before the shift, the opposite was true. "Darkness the norm" evokes William Shatner's remarks, quoted earlier, made following his brief rocket-borne excursion beyond Earth's atmosphere. The same phenomenon was attested more than sixty years earlier by Soviet cosmonaut Yuri Gagarin, the first human in space:

> On the horizon I could see the sharp, contrasting change from the light surface of the earth to the inky blackness of the sky. The earth was gay with a lavish palette of colours. It had a pale blue halo around it. Then this band gradually darkened, becoming turquoise, blue, violet and then coal-black.[11]

The shift to widespread acceptance of black space, according to Brljak, extended over a very long period of time. There are premodern examples of those who conceived of space being dark and late examples of those who continued to think of it as bright, or at least blue. Among Islamic scientists, al-Kindi (c. 800–870) wrote an essay "regarding the cause of the azure-blue color which is seen in the atmosphere in the direction of the

---

[11] Gagarin, *Road to the Stars*, 154.

sky and which is considered to be the color of the sky." In this work he argued that "azure-blue" is not the actual color of the sky "but only something which is exposed to our sight when light and shade encounter it."[12] Al-Biruni (973–after 1050) denied that the night is Earth's shadow.[13] And al-Qarafi (1228–1285) explained that the sky is indeed black. "Under it is the air, which is transparent and luminous. Our gaze penetrates the air and sees it against the sky, so to speak. Thus, the blue color (*lazward*) results from the purity of the air and the darkness of the sky, for we are dealing here with the mixture of the black and the pure."[14] Among Europeans, Leonardo da Vinci (1452–1519) declared that "[t]he blue of air arises from the thick body of illuminated air between the tenebrous [region] above and the earth."[15] And Otto von Guericke (1602–1686) explained that the blue color of the sky by day "has its source in black and white, for at the point where air is devoid of the most rarefied aqueous humors and is absolutely pure, there the white ends and the black begins."[16]

Even after the advent of Copernicanism, however, considerable enthusiasm remained for an interstellar space that was *not* tenebrous, but instead bright. Thomas Digges, while not calling space blue, declared that the Sun "spherically disperses his glorious beams of light through all this sacred celestial temple." Moreover, the starry outer space he envisaged is "garnished with perpetual shining glorious lights innumerable"—which is not very dark-sounding. In the following century, Henry More (1614–1687), a Copernican, repeatedly implied the blueness of space: it is "the azure Orb," "azure skie/. . . from Saturn to the Sunne," "this azure round," "that Circle blue."[17] Traherne with his azure Universe was a contemporary of More. And at least at the popular level, as Brljak shows, the notion of the blue heavens persisted well into the nineteenth century: "Tell us the history of the stars," a poet in 1848 asks of a comet: "Of those bright orbs that shine afar,/In azure realms."[18]

For the scientist, the blackness of space replaced azure realms once it was established that the Universe functioned according to Newtonian physics. Thus air, being affected by gravity, would be held close to Earth. The

[12] Spies, "Al-Kindī's Treatise on the Cause of the Blue Colour of the Sky."
[13] al-Bīrūnī, *The Exhaustive Treatise on Shadows*, 1:22–25.
[14] Hoeppe, *Why the Sky Is Blue*, 41.
[15] Farago, Bell, and Vecce, *The Fabrication of Leonardo da Vinci's "Trattato della pittura,"* 2:151.
[16] von Guericke, *The New (So-Called) Magdeburg Experiments*, 220–21 (4.12).
[17] More, *Democritus Platonissans*, B1ᵛ, B6ʳ, B7ᵛ, C2ᵛ.
[18] Anonymous, "To a Comet," 214–15.

atmosphere would be a thin blanket, not a region extending all the way to the Moon as it did in the two-storey Universe. Under the influence of gravity, moons and planets thus orbited in a *vacuum* that would offer no resistance to halt their motion. Von Guericke, for example, made his remarks about sky color in a work about vacuum experiments.

Since the last half of the twentieth century, thanks to the eyewitness accounts and photographs produced by cosmonauts and astronauts, the blackness of space has come to dominate the popular imagination as well as the scientific mind. And set amid that unfathomable inky backdrop is Earth, our own bright wandering star, a planet like and very unlike other planets in both our own and other Solar Systems. Zooming in on this exquisite blue marble, we ponder its spherical splendor, absent national boundaries, and worth thinking about as "Gaia" in all its rich integrity, coherence, and fragility, like that of a living organism. Zooming out, from a distance of six billion kilometers, as in Voyager 1's famous glimpse of the Pale Blue Dot (to which we'll return), we are perhaps oppressed with a sense of cosmological smallness, loneliness, insignificance, and "existential humility."

Either way, all that immense blackness now renders the stellar character of Earth undeniable. We might even dare to say, as Hans Blumenberg has written of our "cosmic oasis," that Earth—"this miracle of an exception, our own blue planet amid the disappointing celestial desert—is no longer 'also a star,' but rather the only one seeming to deserve that name."[19] No doubt Thomas Traherne would understand.

[19] Blumenberg, *Die Genesis*, 793–94: "Die kosmische Oase, auf der der Mensch lebt, dieses Wunder von Ausnahme, der blaue Eigenplanet inmitten der entäuschenden Himmelswüste, ist nicht mehr 'auch ein Stern,' sondern der einzige, der diesen Namen zu verdienen scheint." English translation adapted from Blumenberg, *The Genesis of the Copernican World*, 685.

7

The Great Copernican Cliché

In 2005 astronomer Mike Brown of Caltech discovered the body now called "Eris" orbiting beyond Pluto. Eris was one of a number of distant bodies, out past Neptune, that astronomers had been discovering then (and are still discovering now). But Eris rivaled Pluto in size. A tenth planet?

No: In 2006, the International Astronomical Union (IAU), meeting in Prague, voted to reclassify Pluto as a "dwarf planet." Eris went into that class, too. So did the large asteroid Ceres.

The result of the IAU's decision was a huge public-relations brouhaha in the media, in some state legislatures, and also among admirers of an astronomer named Clyde Tombaugh (1906–1997), who had discovered Pluto in 1930. Pluto's reclassification was widely viewed as a slight or demotion. Some astronomers ran with this—Brown would go on to write a book with the title *How I Killed Pluto and Why It Had It Coming*. The American Dialect Society subsequently named "plutoed" its word-of-the-year for 2006: "to pluto is to demote or devalue someone or something, as happened to the former planet Pluto." All this reaction concerned mere technical vocabulary as it pertained to a celestial object that no earthly naked eye had ever observed, and that was discovered by Tombaugh only seventy-six years before its "demotion" by the IAU.

How much more of a shock must have resulted, then, when Copernicus demoted the Sun: of all the visible objects in the Universe, the one that for millennia and across countless cultures had been an object of highest honor, in some cases of outright worship. As we noticed in those Ptolemaic and Copernican diagrams with their concentric circles, at one level Copernicus simply swapped around the positions of Earth and Sun. Formerly, Earth was in the center; now the Sun is in the center—and no longer a planet. Plutoed!

When we understand the locational value system of the two-storey pre-Copernican Universe, we grasp what a huge insult to the Sun this might involve; we grasp just how flagrantly the Sun was being plutoed, as indicated by Tolosani's complaint about Copernicus "put[ting] the indestructible Sun

A Universe of Earths. Dennis Danielson and Christopher M. Graney, Oxford University Press.
© Oxford University Press (2025). DOI: 10.1093/9780197803547.003.0007

in a place subject to destruction."[1] Exactly this objection to his cosmology, as we've noticed already, seems to have been Copernicus's motivation for rhetorically "renovating the cosmic basement" and construing the center, in his most highly poetic passages, as a perfectly fitting residence for the Sun. "Behold, in the midst of all resides the Sun. For who, in this beautiful temple, would set this lamp in another or a better place, whence to illuminate all things at once?...Truly indeed does the Sun, as if seated upon a royal throne, govern his family of stars as they circle about him."[2] The very vocabulary of Copernicus's original Latin seems to endorse the fittingness of this arrangement: Of *course* the center is the right place to locate the throne (*solium*) from which the Sun (*sol*) most conveniently rules his stellar family.

If the idea that Copernicus raised the Earth to stardom rather than "demoted" Earth from its supposedly privileged position at the cosmic center seems counter to what you have heard about Copernicus, that might be because Copernicus succeeded in convincing both future astronomers and a wider public that the center was indeed a special place, one befitting a throne—so much so that, from about a century after publication of the *Revolutions*, writers began suggesting that it was Earth that had been *de*throned from its formerly special place in the center—so much so that a certain French king would adopt Copernican Sun imagery for his throne, a move with consequences for Plurality of Worlds, as we shall see. We call this demotion misinterpretation "the Great Copernican Cliché." We need to devote some attention to critiquing it, not only because it *is* a misinterpretation but also because it has become such a standard, almost orthodox take on Copernicus that it routinely gets treated simply as a fact of science history.

Ponder this: Have you ever heard or read that Copernicus *dethroned* or *demoted* humankind by removing Earth from the center of the Universe? Anyone writing on the history of science as it relates to human value seems obliged to say so, including prominent scientists who authoritatively interpret that history for a wider public. In 1973, in one of a series of public lectures marking the five hundredth anniversary of Copernicus's birth, Theodosius Dobzhansky declared that, with Copernicus, Earth was "dethroned from its presumed centrality and pre-eminence."[3] Carl Sagan

---

[1] Tolosani, "Tolosani's Condemnations of Copernicus' *Revolutions*," 189.

[2] Copernicus, *De revolutionibus*, I.x; translation from *BOTC* [*The Book of the Cosmos: Imagining the Universe from Heraclitus to Hawking*, ed. Danielson], 117.

[3] Dobzhansky, *Man's Place in the Universe*, 80. Compare Jürgen Hamel, *Nicolaus Copernicus: Leben, Werk und Wirkung*, 300: "Die 'Entthronung' des Menschen, die mit der Verdrängung der

described Copernicanism as the first in a series of "Great Demotions . . . delivered to human pride."[4] And the same general claim continues to be repeated year by year, whether in popular accounts or in the writings of the most learned scientists, as for example in the pronouncement of England's Astronomer Royal, Sir Martin Rees: "It is over 400 years since Copernicus dethroned the Earth from the privileged position that Ptolemy's cosmology accorded it."[5] More recently, Caleb Scharf has written an entire book that builds upon such claims: Copernicus's *Revolutions* offered a "completed heliocentric model of the universe, shifting the Earth from the center of the cosmos to a secondary place, spinning and orbiting around the Sun— a demotion that would reshape our species' scientific history."[6] By contrast, we're arguing that it was a promotion—and indeed it helped humankind imagine that the whole Universe might be full of other earths with other humankinds on them. No one thinks of other earths as dethroned, plutoed places; plutoed places are not where one boldly goes, seeking new life and new civilizations.

One feature of references such as Scharf's to Earth's and our alleged dethronement is that they are left unargued, stated as something merely obvious, at least to us enlightened moderns. To his credit, Scharf notices that it *wasn't* obvious to some of the most prominent early Copernicans: "it doesn't appear that either Galileo or Kepler saw heliocentrism as a *demotion* of the terrestrial status. Quite the contrary: it meant we were no longer at the 'bottom' of the planetary pile."[7] Yet this astute insight seems not to deflect Scharf or most other commentators from a determined reading of Copernicanism as entailing, if one really understood its consequences, Earth's and humankind's cosmic demotion.

We're calling the "terrestrial demotion interpretation" of Copernicus a cliché to highlight its sheer pervasiveness in accounts of the history of astronomy. Meme-like, it has been repeated so often, and by such respectable voices, that it now seems virtually a part of everyone's mental furniture. Such is the nature of a cliché: Its very frequency of repetition results, independent of its truth or falsehood, in its being repeated yet again. Perhaps not every cliché is false, but this one is.

Erde aus der Weltmitte erfolgte, war erst der Beginn der Relativierung der Stellung des Menschen im Kosmos."
   [4] Sagan, *Pale Blue Dot*, 26.
   [5] Rees, *Before the Beginning*, 100.
   [6] Scharf, *The Copernicus Complex*, 4.
   [7] Scharf, *The Copernicus Complex*, 23–24.

A quick look at Islamic, Jewish, and Christian thought, not to mention common language, helps drive home the falseness of the idea that the downstairs in a two-storey Universe, the center, was a good place to be. *Upward* is the direction of improvement and rising importance. Within these religions, Heaven is *up*; the spirits of the devout are *exalted*—literally, "lifted high"—and so on. By contrast, *downward*, toward the center, is the direction of deterioration, corruption, the grave, and hell. In this sense, as Martianus Capella (fl. 410–439) pointed out in his cosmological writings, Earth is "in the middle *and at the bottom*" position in the Universe.[8] As the geographer Al-Biruni (973–1048) stated, "in the centre of the sphere of the moon is the earth, and this centre is in reality the lowest part."[9] Thomas Aquinas, the greatest of medieval Christian philosophers, declared that, "in the universe, earth—that all the spheres encircle and that, as for place, lies in the center—is the most material and coarsest (*ignobilissima*) of all bodies."[10] Moreover, based on a consistent extrapolation from this view, the Middle Ages conceived of hell as being located at the *very* center and therefore coincident with the center of Earth. In Dante's *Divine Comedy*, accordingly, there we find the Inferno, hell itself. At Earth's inmost core, at its very center, at the very lowest place in the Universe, in keeping with Aristotelian physics as well as with poetic justice, appears Satan. And he is not dancing in flames—for the element of fire tends heavenward—but frozen, immobile, in ice.

Thus pre-Copernican cosmology pointed to the sheer grossness of humankind and its abode. In this view, Earth appears as a universal pit, figuratively as well as literally the world's low point. This negative view encompasses, finally, not only ancient and medieval Muslim, Jewish, and Christian writers, but also many prominent voices that we usually associate with Renaissance humanism, both before and after the time of Copernicus. We have already encountered Giovanni Pico's reference to our Earth in terms of excrement and filth (this in a work that acquired the title *Oration on the Dignity of Man*—1486). And a quarter century after the publication of *Revolutions*, in 1568, Michel de Montaigne takes up the same theme once more, declaring that we are "lodged here in the dirt and filth of the world, nailed and rivetted to the worst and deadest part of the universe, in the lowest storey of the house, and most remote from the heavenly arch."[11]

---

[8] Capella, *The Marriage of Philology and Mercury Capella*, 318 (emphasis added).

[9] Al-Biruni, *The Book of Instruction in the Elements of the Art of Astrology*, 45.

[10] Thomas Aquinas, *Commentary on Aristotle's De Caelo* (1272 s.), II, xiii, 1 & xx, n. 7, in vol. 3, p. 202b of the Leonina ed.; trans. (and quoted) by Brague, "Geocentrism as a Humiliation for Man," 202.

[11] Montaigne, "An Apology of Raymond Sebond," 2:134.

How surprised might Pico and Montaigne—or Ptolemy, for that matter—be to read confident declarations by later leading scholars as well as scientists to the effect that Earth's removal from that "lowest storey" constituted a severe demotion. How surprised might Galileo be? In his 1632 *Dialogue Concerning the Two Chief World Systems: Ptolemaic and Copernican*, his spokesman Salviati declares: "As for the earth, we seek . . . to ennoble and perfect it when we strive to make it like the celestial bodies, and, as it were, place it in heaven, from whence your philosophers have banished it."[12] "Your philosophers," in this case, of course, are the sorts of Ptolemaic astronomers who, according to the almost unanimous account of historians of science for at least the past century, placed Earth "on a pedestal" at the center of the world. Salviati knows that it is *helio*centrism, the new cosmology of Copernicus, that raises up on a pedestal the abode of humankind, to be a star like Mars and Jupiter. If in Ptolemaic cosmology, the place of our Earth is both low and lowly, then in the cosmology of Copernicus and Galileo it is, in more senses than one, *uppity*.

This uppityness matters for our story of our Earth and other worlds, from Copernicus to NASA. We want to reach for the stars. Who would care about, let alone seek to build ships to traverse, a Universe full of the worst and deadest? Going boldly implies destinations worth going to.

But if all this seems counter to what you have heard, that is because what Copernicans and pre-Copernicans actually wrote does not square with the pronouncements of more modern commentators. The latter make repeated claims about that "outrage" against humankind's "naive self-love" which we associate with "the name of Copernicus."[13] (These are the words of Sigmund Freud, who liked fancying himself as a "Copernicus" in his own domain of science.) Actual Copernican doctrine does not harmonize with the same tale as told more recently by Sagan or by anthropologist Terrence Deacon, who writes that "Since Copernicus first suggested that Terra Firma might not be located in the center of the cosmos, most of the remaining vestiges of human specialness have come into doubt."[14] What Copernicans and pre-Copernicans actually wrote undercuts the fundamental assumption of such pronouncements, that central location equates with human specialness, and that Earth's loss of central location equates with a denial of human specialness.

---

[12] Galileo, *Dialogue*, 42.
[13] Freud, *A General Introduction to Psycho-Analysis*, 252.
[14] Deacon, "Giving up the Ghost," 16–17.

So, if prominent medieval authors and authoritative early modern spokespeople for heliocentrism undermine that equation of geocentrism and human specialness—and if Copernicanism creates the exhilarating prospect of our species inhabiting a star, a planet, a place no longer "excluded from the dance of the stars"—then how did the Great Copernican Cliché arise? How did we get to that story you have heard, about Copernicus demoting Earth (and by extension, us human beings)?

Although we have our suspicions, we can't offer a definitive answer. It is much easier to expose factual error than to account for motivation. Undoubtedly, for many, the exhilaration of a new Universe translated into bewilderment. One thinks again of John Donne's oft-quoted lament, "'Tis all in pieces, all coherence gone"; or Pascal's "The eternal silence of these infinite spaces frightens me"; or Robert Burton's humorous but frustrated roundup of the cosmologists of his day: "the world is tossed in a blanket amongst them, they hoist the earth up and down like a ball, make her stand and go at their pleasures."[15] And perhaps such bewilderment and loss of security have been interpreted, understandably, as loss of specialness.[16] But it is worth pointing out that the security, even coziness of the two-storey Universe (if "secure" and "cozy" apply to a Cosmos in which the Earth is a vanishingly small turd ball at the center of a vast ensemble of eternally perfect ethereal machinery), does not justify ignoring the written record and reading human haughtiness or vanity into that cosmology.

However, by the mid-1650s—more than a century after publication of Copernicus's *Revolutions*—we do find some writers associating geocentrism with human self-importance. Among these are Cyrano de Bergerac, who protests "the insufferable pride of humans," and Thomas Burnet, who takes delight in referring to our Earth as an "obscure and sordid particle."[17] But it is the great French popularizer of Copernicanism Bernard Le Bovier de Fontenelle (1657–1757; those dates are correct) who most powerfully asserts the negative axiological implications of the new cosmology. As we will see, his 1686 book *Conversations on the Plurality of Worlds* would give a great boost to the idea of a Universe of other suns and other earths. It was written in the form of a discussion between a narrator and a Countess. In it, the

[15] Donne, *An Anatomy of the World*; Pascal, *Pensées*, 78; Burton, *The Anatomy of Melancholy*, 329.
[16] It is reasonable to suppose that the Plurality of Worlds debate and the "infinitizing" of the Universe in writers as diverse as Bruno and Newton played a major role in occasioning this kind of anxiety.
[17] Bergerac, *Les états et Empire de la lune* (Paris, 1656), and Burnet, *Telluris Theoria sacra* (London, 1681), both quoted by Paolo Rossi, "Nobility of Man and Plurality of Worlds," 151, 155.

Countess, upon hearing about the heliocentric model, declares disapprovingly that Copernicus, had he been able, would have deprived Earth of the Moon just as he has deprived it of all the other planets; for she perceives, she says, that he "had no great kindness for the earth." Yet Fontenelle's own character replies to the contrary by praising Copernicus: "It was well done of him . . . to abate the vanity of mankind, who had taken up the best place in the universe."[18] This interpretation of Copernicanism became the standard and apparently unquestioned version of the Enlightenment, as magisterially summarized by Johann Wolfgang Goethe in 1810:

> Perhaps no discovery or opinion ever produced a greater effect on the human spirit than did the teaching of Copernicus. No sooner was the Earth recognized as being round and self-contained, than it was obliged to relinquish the colossal privilege of being the center of the Universe.[19]

From Goethe and the Enlightenment to the present there has been, in more senses than one, almost no looking back. How might we further account for the genesis of this interpretation and for its manifest success in driving out all others? We have already mentioned the contribution made by Copernicus in "renovating the cosmic basement," whose success made the central location to appear indeed as a very special place, and perhaps allowed us, anachronistically, to read the physical center's post-Copernican excellence back into the pre-Copernican world picture—and so turn things upside down. In some respects, this may be no more than just an innocent confusion. Less innocently, however, the Great Copernican Cliché may function as a self-congratulatory story that serves to displace our own modern hubris onto "the Dark Ages" and onto anyone else who might not know that Earth orbits the Sun (non-European or Indigenous cultures, perhaps). When Fontenelle and his successors tell the tale, it is clear that they are making no disinterested point (one thinks again of Freud); they make no secret of the fact that in their misanthropic way they are "extremely pleased" with the demotion they read into the accomplishment of Copernicus. But the trick of this supposed dethronement is that, while purportedly rendering

---

[18] Fontenelle, *Conversations*, 25–26.
[19] Goethe, "Materialien zur Geschichte der Farbenlehre," 81: "Doch unter allen Entdeckungen und Überzeugungen möchte nichts eine größere Wirkung auf den menschlichen Geist hervorgebracht haben, als die Lehre des Kopernikus. Kaum war die Welt als rund anerkannt und in sich selbst abgeschlossen, so sollte sie auf das ungeheure Vorrecht Verzicht tun, der Mittelpunkt des Weltalls zu sein."

"humankind" less cosmically important, it enthrones modern "scientific" humans in all our enlightened superiority. It declares, in effect, "We're truly very special because we've shown that we're not so special." Those without modern science, be they in the distant past or not, lack that superiority.

By equating human specialness with the now indefensible belief in geocentrism, such modern ideology manages to treat as nugatory or naive the legitimate and burning question of whether Earth or Earth's inhabitants may indeed be in some sense cosmically special. Instead, it offers, if it offers anything, a specialness that is cast in exclusively existential or Promethean terms, with scientific humankind lifting itself up by its own bootstraps and heroically, though in the end pointlessly, defying the universal silence. But such suppression or evasion of the larger questions of teleology ought to be recognized as lacking historical, philosophical, or scientific warrant.

Not that everyone has evaded such questions. There appear to be an increasing number who unapologetically carry the torch for this "rare earth," to echo one prominent contribution on this theme.[20] And for us today (as they would have for Kepler), the words of Michael Polanyi written in 1958 still strike a chord: Scientific "objectivity" as "exemplified by Copernican theory . . . does not demand that we estimate man's significance in the universe by the minute size of his body, by the brevity of his past history or his probable future career. It does not require that we see ourselves as a mere grain of sand in a million Saharas. . . . It is not a counsel of self-effacement, but the very reverse—a call to the Pygmalion in the mind of man."[21] Such an account of scientific objectivity, this Pygmalion in the mind, may in the end prove compatible with what some physicists and some nonphysicists would see as a quest for meaning that transcends the physical. But even at the physical level, there's insufficient reason to see Copernicanism's reinterpretation of Earth as a wandering star as a demotion or dethronement.

So, what you have heard about Copernicus plutoing the Earth is wrong. Copernicus made Earth a star. What he plutoed was the Sun. It seems he did such a good job, however, of renovating the Sun's "plutoed" position that we all forgot just what it is he did.[22]

---

[20] Ward and Brownlee, *Rare Earth.*

[21] Polanyi, *Personal Knowledge,* 5.

[22] A longer, earlier version of this chapter appeared as D. Danielson, "The Great Copernican Cliché," *American Journal of Physics,* 69.10 (Oct. 2001): 1029–35 (with the permission of AIP Publishing).

8

The Case Against Copernicus

What Copernicus did with the Earth and the Sun is hardly the only forgotten thing in the history of science. We have also largely forgotten the scientific case against Copernicus and against some of the ideas of Copernicans like Galileo. We have forgotten that the edge of the Moon is seen to glow.

The painting "Two Men Contemplating the Moon" by the eighteenth-century artist Caspar David Friedrich shows this clearly (see Figure 8.1). Friedrich paints the crescent Moon, with its dark side illuminated by what Galileo identified as Earthshine, but with the edge of that dark side distinctly bright. Even today this edge glow is regularly seen and represented by people drawing and painting the Moon.

Locher and Scheiner rejected Galileo's idea that the Earth illuminated the Moon.[1] They cited the edge glow. The source of the dark side illumination, they postulated in their 1614 *Disquisitions*, is sunlight penetrating through the body of the Moon. The edge glow is akin to what is seen in back-illuminated clouds or crystals, they said. Contrary to "certain recent ideas," the appearance of a crescent Moon is consistent with its being a dark but translucent body—"neither completely diaphanous, nor entirely opaque"— that both reflects the Sun's light from its surface (the illuminated crescent) and transmits the Sun's light through itself (the dark side illumination). The idea of the Moon as translucent is not new, they pointed out, citing sources dating back to the fifth-century writer Macrobius.

Locher and Scheiner were telescopic astronomers. They saw everything on the Moon that Galileo saw, spots, mountains, and all, and they showed it all in a page-sized drawing they included in *Disquisitions* (see Figure 8.2). That drawing also showed the edge glow, a glow that they said was often enhanced by a "sprinkling of luminous points" within it. These had been seen, they said, by a variety of observers with even average-quality telescopes.

---

[1] All Locher and Scheiner quotations here from "Disquisition 27," in *MathDisq* [Locher with Scheiner, *Disquisitiones Mathematicae*, tr. Graney], esp. 69, 72.

Figure 8.1 Central portion of "Two Men Contemplating the Moon" by the eighteenth-century artist Caspar David Friedrich (1819 version). Note the glow of the "dark" edge of the Moon, opposite the illuminated crescent.
Credit: Wikimedia Commons

To their minds, the telescope did not tear down the two-storey Universe. "Earth is not a star," they explicitly state; it shines too feebly. "The moon is not an Earth"; the Moon and the Earth are very different bodies. They also mention, interestingly enough—and correctly—that the Moon is not habitable, and not the home to life. No, what the telescope did, according to them, was show that the two-storey Universe was different, and more interesting than previously imagined. It was so interesting that Locher and Scheiner would, in *Disquisitions*, illustrate all the things the telescope revealed, and that Scheiner would go on to dedicate years just to the telescopic study of the Sun.

Today we know that the Universe is not two-storey in nature. The Moon, stars, and everything else we see out there have been found to be made of the same substances that make up our Earth and to obey the same laws of physics that govern things here on Earth. So we tend to think that science supported those like Copernicus and Galileo who rejected the old Aristotelian ideas

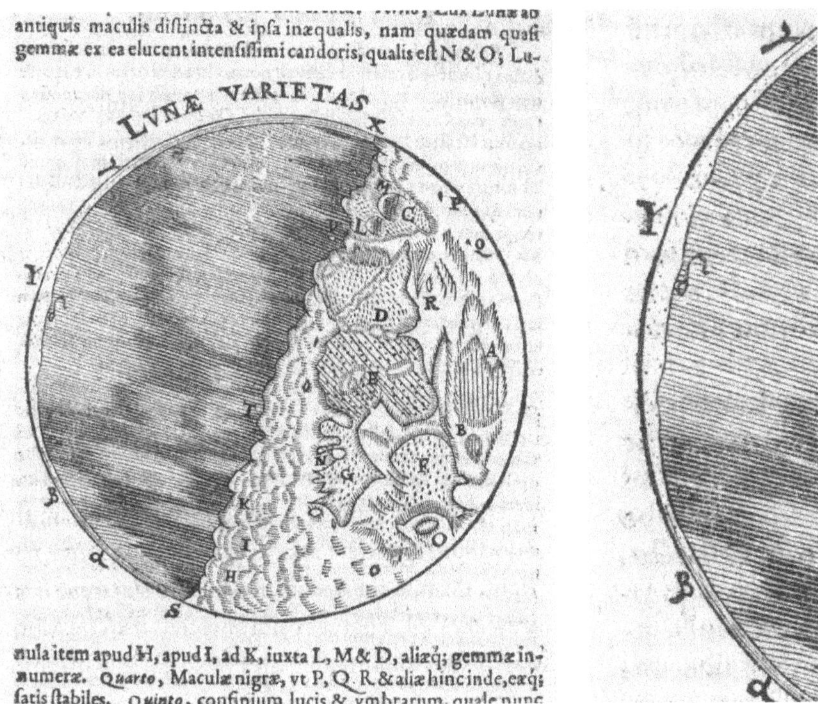

Figure 8.2 Locher and Scheiner's 1614 sketch of the Moon. Note the illustration of "edge glow" on the left edge of the Moon (detail at right). ETH-Bibliothek Zürich.

about the Universe. And we perhaps tend to think of those who defended the two-storey Universe in the manner that the modern astrophysicist and bestselling author Mario Livio has bluntly characterized them—as *science deniers*, people who reject "painstaking scientific observations and brilliant deductions, only because they *appeared* to contradict some sacred, ancient, vague, poetic texts."[2]

But Livio's caricature is unfair and misleading. Locher and Scheiner opposed Galileo's ideas about the Moon because of their own painstaking scientific observations and brilliant deductions. The edge glow is a real phenomenon. Today we know that it eludes photography and seems to be an optical illusion[3] (all unknown in the 1610s). Indeed, Scheiner with his

---

[2] Livio, *Galileo and the Science Deniers*, 115.

[3] As has been recently discussed by *Sky & Telescope* magazine's Bob King. See King, "Solving an Earthshine Mystery."

study of the Sun would redefine "painstaking scientific observations," yet he continued to adhere to the two-storey Universe.

A fair account must acknowledge that there were a good number of scientific reasons for sticking with the two-storey Universe besides the edge glow of the Moon. For example, physics. If the Moon is a body like Earth, what makes it move? What keeps it up in the sky? Earthy stuff, like rock, is heavy and hard to move. It falls downward if not supported. Livio notes how Cesare Cremonini, a colleague of Galileo's who was both an Aristotelian philosopher and an atheist, had no interest in Galileo's lunar observations: the Moon had not fallen; it was therefore not Earth-like. Period.[4]

Decades before Isaac Newton was even born, there was no physics to explain how an Earth-like, *and therefore heavy*, Moon would circle Earth, or how the Earth itself, an immensely heavy ball of rock and water, would circle the Sun. By contrast, in a two-storey Universe, the ethereal bodies in the upper storey, like the Moon or Sun, could be assumed merely to move naturally, or to be incredibly light, despite their size. They were *different* from things here in the lower storey, after all. That was simply obvious. The Moon had been going around the Earth for all human history. No earthly machine could keep running forever in such a manner.

Another physics argument for the two-storey Universe with its immobile Earth was the motion of falling bodies and projectiles like cannonballs. If Earth were rotating, then different points on its surface would be moving at different speeds. The ground at the poles would merely turn in place, whereas the ground on the equator would move at over 1000 mph (since the Earth, 25,000 miles in circumference, would turn around once in 24 hours). And the top of a mountain at the equator would trace out a bigger circle, and thus move faster, than its base. So would a rock dropped from the top of a cliff at the equator deflect sideways as it fell, thanks to the excess speed it had at the top? Would it *not* fall straight down?

Various anti-Copernicans, from Tycho Brahe to Locher and Scheiner, to (later in the seventeenth century) the Jesuit astronomers Giovanni Battista Riccioli and Claude François Milliet Dechales, would seize on this idea and argue that falling bodies and cannonballs would have to deflect from their expected paths if Earth rotated. No such deflection was detected, and that was an argument in favor of an immobile Earth.

---

[4] Livio, *Galileo and the Science Deniers*, 98.

In fact, they do deflect. The phenomenon is now called the *Coriolis Effect*, and meteorology students study it as the source of the rotation in hurricanes. But it turns out to be surprisingly difficult to detect in cannonballs and dropped rocks, thanks to things not understood in the seventeenth century like the turbulence that forms in the air flowing around a rapidly moving body. Dechales got the science of the Coriolis Effect right, producing illustrations and discussions that could serve in a modern meteorology textbook (see Figure 8.3). But since what he was illustrating had never been seen, Dechales used his science against Copernicus.

The most powerful argument against Copernicus, however, involved not the Moon or the Earth, but rather the stars. It would, and does, have repercussions for the entire idea of other earths, for the idea of *Star Trek* and the Marvel Cinematic Universe. Its roots reached back to an ancient debate about science and the Book of Genesis.

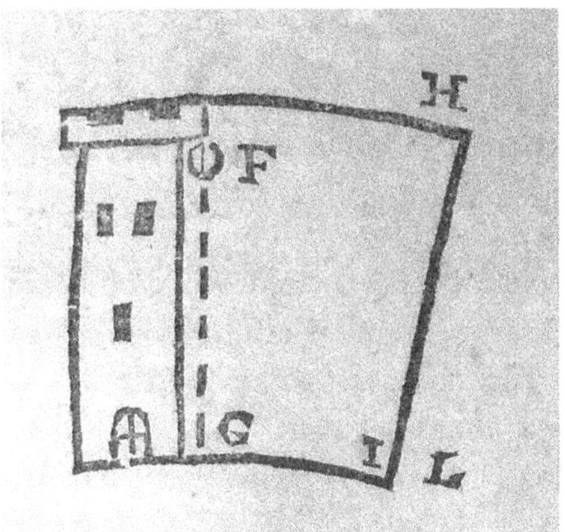
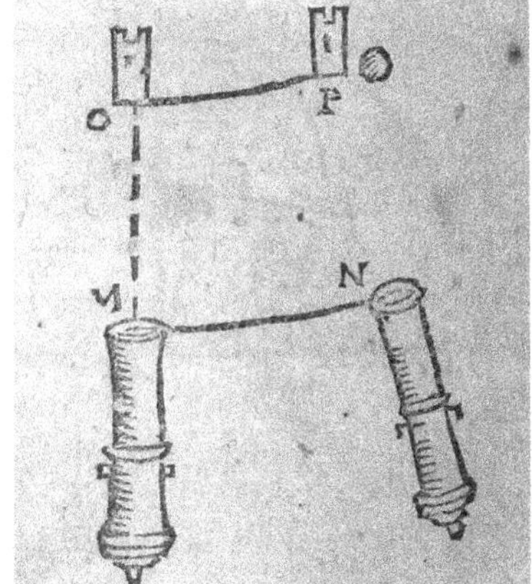

Figure 8.3 Illustrations by DeChales (1674) of what today is called the *Coriolis Effect*. Left: A heavy ball (F) dropped from a tower on a rotating Earth should deflect as it falls because while the ball falls the top of the tower moves from F to H on account of Earth's rotation, whereas the bottom moves a lesser distance, from G to I. Thus, the ball, which was moving at the speed of the top of the tower, ends up at L, away from the base of the tower. Right: For similar reasons, a cannonball launched northward at a target deflects to the right, missing the target because the ball moves rightward at the speed of the cannon (which travels from M to N owing to Earth's rotation), which is greater than the speed of the target (which travels from O to P).
Credit: Vatican Observatory Library.

Genesis 1:14–16 describes the creation of the Sun, Moon, and stars, and in doing so alludes to their sizes: "And God made two great lights; the greater light to rule the day, and a lesser light to rule the night: *he made* the stars also" (King James Version). If the sky were a dome with these lights on its surface, then their "greatness" would be a matter of simple sight. The Sun, Moon, and stars would all be the same distance from Earth; therefore the relative physical sizes of these bodies would simply be what the eye sees: the Sun and Moon would indeed be largest, "greater" than the stars in terms of actual physical bulk.

Careful study of the sky, however, reveals it is not a dome. Ptolemy in his *Almagest* discussed how the appearance of the stars does not depend on the place on Earth from which one observes them. This was another argument showing that the size of the Earth is like a point compared to the distance to the stars:

> Now, that the earth has sensibly the ratio of a point to its distance from the sphere of the so-called fixed stars gets great support from the fact that in all parts of the earth the sizes and angular distances of the stars at the same times appear everywhere equal and alike, for the observations of the same stars in the different latitudes are not found to differ in the least.[5]

The stars of Orion's belt, for example, look no different when observed from Ethiopia than they do when observed from Moscow. As we saw in Chapter 3 and Figure 3.1, such observations contrast with those of the Moon, which reveal that Earth is *not* merely a point in comparison to Earth's distance from the Moon. The Moon shows an obvious parallax; the stars do not.

Ptolemy's science was persuasive. Anyone who traveled and had good eyesight could observe the constancy of the stars, which testified to their vast distances. Thus, despite the apparent contradiction between Ptolemy's Cosmos and the domed Universe described by Genesis, the Roman Christian writer Severinus Boethius (c. 480–524), in his *On the Consolation of Philosophy*, would affirm Ptolemy and write:

> You have learned from astronomy, that this globe of earth is but as a point, in respect to the vast extent of the heavens; that is, the immensity of the celestial sphere is such that ours, when compared with it, is as nothing, and vanishes.[6]

---

[5] Ptolemy, "Almagest" I:6, 10.
[6] Boethius, *Consolation of Philosophy*, 67.

Ptolemy also determined that the stars have small but measurable apparent sizes. Their apparent diameters, he calculated, were a couple of *minutes of arc*—roughly one-fifteenth the apparent diameter of the Moon. The vast distance to the stars meant that they actually had to be very large to appear even that small in the night sky. Ptolemy further calculated the actual diameter of the most prominent stars to be more than four times that of Earth. He calculated that the Sun was actually five times Earth's diameter and thus twenty-five times its bulk, while the Moon was actually less than one-third of Earth's diameter[7] (see Figure 8.4).

A prominent star was therefore far greater than the Moon. Indeed, every visible star in the night sky would greatly exceed the Moon in terms of bulk. Ptolemy's calculations showed that the Moon hardly qualified as one of the great heavenly lights. Anyone with good eyesight who cared to look could at least approximately confirm Ptolemy's measurements of the apparent sizes of the stars compared to the apparent size of the Moon. The stars might *appear* small, but the Moon *was* small.

This apparent conflict with Genesis brought the star-size issue to the attention of Augustine of Hippo (354–430), the African Bishop who was a contemporary of Macrobius. Augustine discussed the issue of star sizes and distances and Genesis in his *On the Literal Interpretation of Genesis*: "Many of the stars, ... [astronomers] boldly assert, are equal to the sun, or even greater, but they seem small because they have been set further away."[8]

Figure 8.4 Relative bulks of the Moon (left) and
a prominent star (right) according to Ptolemy.

[7] Van Helden, *Measuring the Universe*, 27.
[8] Augustine, "The Literal Meaning of Genesis," 211 (Book II, 16.33).

Augustine's view regarding the "two great lights" was that they are "great" to our eyes: "Grant this to our eyes, after all, that it is obvious that they [Sun and Moon] shine more brightly than the rest upon the earth, and that it is only the light of the sun that makes the day bright, and that even with so many stars appearing, the night is never as light when there is no moon, as when it is being illuminated by its presence."[9]

That the stars were larger than the Moon, despite Genesis, was widely acknowledged by Christian interpreters across a millennium and a half. Thomas Aquinas (c. 1225–1274) addressed the "great lights" question in his *Summa Theologica*, concluding like Augustine that "as the senses are concerned, its [the Moon's] apparent size is greater."[10] John Calvin (1509–1564) made the same general point as Augustine and Aquinas. Calvin praised astronomy as "pleasant," as "useful," and as "unfold[ing] the admirable wisdom of God."[11] The Holy Spirit "had no intention to teach astronomy," he wrote, and so while astronomers may understand that the Moon is actually a comparably small astronomical body, "the Holy Spirit would rather speak childishly than unintelligibly to the humble and unlearned"[12] who every night see for themselves the Moon appearing much larger than any star.

Thus, it was widely understood that stars were much bigger than the Moon. But if Earth circles the Sun, as Copernicus proposed, then Earth's *orbit* replaces Earth's *globe* in Ptolemy's argument about the distance to the stars. Observations of the same stars from different locations on Earth's *orbit* (that is, at different times of the year) are not found to differ in the least. There is no *annual* parallax. Thus, Earth's *orbit* is like a point compared to the distance to the stars, as we've noted in previous chapters.

But it therefore follows that if, under Ptolemy, a prominent star measured perhaps four times the diameter of *Earth* (and thus a little smaller than the Sun), then under Copernicus that star becomes four times the diameter of *Earth's orbit*. That prominent star now utterly dwarfs the Sun. For that matter, so does every star seen in the night sky, even the least prominent. Under Ptolemy, stars were consistent in size with the Sun and planets (see Figure 8.5). Under Copernicus, they became an entirely different class

---

[9] Augustine, "The Literal Meaning of Genesis," 212 (Book II, 16.33).

[10] Aquinas, "*The Summa Theologica*," 242 (Question LXX: "Of the Work of Adornment, as regards the Fourth Day—In Three Articles").

[11] Calvin, *Commentaries on the First Book of Moses*, Chapter 1, par. 16, 86–87.

[12] Calvin, *Commentary on the Book of Psalms*, "Psalm CXXXVI," par. 7, 184–85.

Figure 8.5 Illustration from the *Harmonia Macrocosmica* of Andreas Cellarius (1660), showing the relative sizes of the Sun, stars, and planets in a geocentric Cosmos. The Sun is the largest circle shown. Everything else is a measurable fraction of the Sun's size. Under a Copernican cosmos, the Sun and planets would be reduced to dots compared to the stars.
Credit: ETH-Bibliothek Zürich.

of bodies, gargantuan and monstrous. Recall Kepler's calculation that Sirius had to be larger than the orbit of Saturn.

On solid observational and mathematical grounds, therefore, Copernicanism seemed to imply something completely ridiculous. And accordingly, a rejection of heliocentrism, in light of the best evidence available at the time, was a far cry from "science denial."

Astronomers faced a real dilemma. True: Copernicus freed his followers from having to believe the likewise ridiculous claim that the entire sphere of the fixed stars rotates about Earth once every twenty-four hours—a well-known and controversial feature of the two-storey Universe. In *Paradise Lost* (1667, 1674), Milton's Adam expresses astonishment that, seemingly, the "stars roll/Spaces incomprehensible" every day, thus exhibiting "incorporeal speed . . . to describe whose swiftness number fails" (7.15–38). Yet the Copernican alternative to accepting *that* proposition was to accept the crazy conclusion that every star was bigger than the size of the whole *orbis magnus*: Earth's orbit. This was Scylla and Charybdis: two monsters with no apparent safe path between them. Each one seemed to involve an absurdity.

Early committed Copernicans therefore simply had to embrace gargantuan stars. Recall that Digges's *Perfit Description* (1576) described stars in a Copernican Universe as extending infinitely outward and "far excelling our sun both in quantity and quality." In a Ptolemaic Universe, the stars outnumbered the Sun, of course, far excelling it in *numerical* quantity, perhaps, but they did not beat out the Sun in any other way. Digges recognized that in a Copernican Universe, the stars themselves far excel the Sun.

Digges means "size" when he speaks of "quantity." When he discusses how something the size of Earth's orbit could be vanishingly small in a Copernican Universe, he writes, "every quantity [i.e., size] has a certain proportionable distance whereunto it may be discerned, and beyond the same it may not be seen."[13] The word "quantity" was often used to mean, as one writer at the turn of the seventeenth century put it, "the magnitude of the bigness"; thus, another writer could say that "bodies . . . without quantity are no bodies."[14] For Digges, the stars "far excel the sun" in *size* and quality—as is fitting for a starry Universe that to him was "the palace of felicity," "the court of celestial angels," and "the habitacle for the elect."

Other Copernicans also saw the star size difficulty. They just swallowed hard and accepted it. In 1629, in the age of the telescope, Philips Lansbergen, writing in his native Dutch, described the immense Copernican stars as giant living beings, God's own army and the palace guard of heaven itself. Their hugeness was fitting because it testified to the power of God. Lansbergen found plenty of scriptural verses to support his thinking about giant Copernican stars, such as the Canticle of Deborah and Barack, where "the stars in their courses" fight a battle (Judges 5:20).[15]

But many astronomers gagged at having to embrace the giant stars of the Copernican Universe. Tycho Brahe, in laying out the case that the Copernican Universe required giant stars, cited them as an absurdity.[16]

---

[13] Digges, *A perfit description*, sig. N.3.ᵛ.

[14] Bell, *The Jesuites Antepast*, 55; Parsons, *A Review of Ten Publike Disputations*, 220–21. Both authors are engaged in disputes with Catholics regarding the Eucharist that involved questions of size and divisibility. See Parsons, 167–70, for more equation of "quantity" with size magnitude.

[15] Graney, *Setting Aside*, 76–85.

[16] Dreyer, *Tychonis Brahe Dani: Epistolæ*, 197. Dreyer, *Tycho Brahe*, 176–77.

Even Sagredo, the intelligent "neutral" character in the debate between the pro-Copernican Salviati and the anti-Copernican Simplicio in Galileo's 1632 *Dialogue*, agreed. Sagredo acknowledges the anti-Copernican parallax-based argument about the distances required for stars in the Copernican Universe, distances that would make the sphere of stars "so immense that in order for a fixed star to look as large as it does, it would actually have to be so immense in bulk as to exceed the earth's orbit—a thing which is, as they say, entirely unbelievable."[17]

Today historians of science understand that the Copernican system required the starry Universe to be vastly larger than had been previously imagined. That conclusion fits with our modern understanding of the Universe as extending unimaginably far beyond the bounds of our Solar System. But it is widely forgotten that the Copernican Universe required the *stars themselves* to be vast and that this requirement presented a genuine scientific impediment to the acceptance of heliocentrism. Again, this wasn't simply a matter of science denial.

Our modern view of stars is certainly not that they are a completely separate class of bodies, with every star visible in the night sky dwarfing the Sun. The apparent sizes of stars that Ptolemy and many astronomers after him determined—the "one fifteenth the apparent diameter of the moon" estimate mentioned earlier—turn out to be illusory, like the edge glow of the Moon. The sizes of stars that we see with our eyes are tricks of light; the stars are actually far smaller than they appear.

Even when a seventeenth-century telescope was used, that illusion remained (see Figure 8.6). The apparent size of the star was certainly *modified* by the telescope. Galileo spoke of his as removing the glare from a star (here he meant either a fixed star or a wandering star/planet) and as "showing the disc of the star bare and very many times enlarged."[18] In the case of Venus, the telescope clearly revealed that what the naked eye saw was illusory: Venus appeared much smaller (compared to the Moon) than it did when seen by the unaided eye; and while the eye always saw Venus as a dot, the telescope revealed a changing, sometimes even crescent, form. The telescope showed the fixed stars to be likewise smaller than what the naked eye saw but still displaying measurable discs. Galileo reported their diameter to be, in the case of brighter stars, not 1/15 the apparent diameter of the Moon, contrary to Ptolemy, but 1/360 of it.[19]

---

[17] Galileo, *Dialogue*, 432.
[18] Galileo *Dialogue*, 418–19.
[19] 5 seconds of arc. Galileo, *Dialogue*, 417.

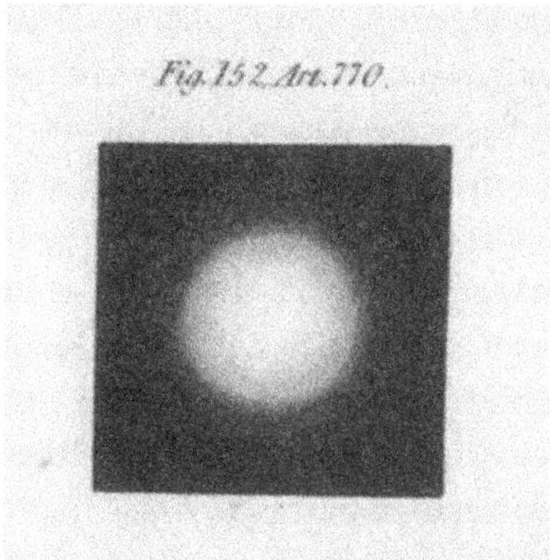

Figure 8.6 Illustration of a star as seen through a small-aperture telescope such as was used in the seventeenth century, from John Herschel's nineteenth-century *Treatises on Physical Astronomy, Light and Sound*. The globe-like appearance is a product of the optics of the telescope and does not reflect the size of the star; the star's appearance is vastly inflated.
Credit: ETH-Bibliothek Zürich.

Today we know that Venus and the fixed stars are not the same. The telescope reveals Venus truly, but in the case of stars, the telescope succumbs to the same trick of light as the eye. We know now that even the 1/360 measurement is vastly too large.

Yet that measurement is what telescopes of the time consistently revealed to careful observers. Throughout much of the seventeenth century, such measurements were reproducible by anyone with a telescope. They were not understood to be illusory but were considered real and valid. They were solid science.

Seventeenth-century anti-Copernicans wielded star measurements deftly. Locher and Scheiner kept the argument simple and brief: In the Copernican Universe, the Earth's orbit is *as a point*, vanishingly small; the stars, appearing small but measurable, are larger than points; therefore, it logically follows that in a Copernican Universe, all visible stars must be larger than the Earth's orbit (and must accordingly dwarf the Sun); therefore, the Copernican system is to be rejected. The Jesuit André Tacquet likewise kept it simple and brief, developing the analogy between Earth in the Ptolemaic system and Earth's orbit in the Copernican system that we used above.[20]

---

[20] *MathDisq*, xxii-xxiii, 30; Graney, "Galileo Between Jesuits."

Giovanni Battista Riccioli (1598–1671) had no interest in simplicity or brevity. He inundated the readers of his 1500-page, un-humbly titled *New Almagest* with tables of data on measurements of star diameters, with calculated actual sizes of stars under various models, and so on—and he showed the same thing: Under Copernicus, stars would have to be huge. "I just want this information to be available," he smirked, "in case anyone maybe feels like discussing, say, the hypothesis of Copernicus." The only recourse for the Copernicans, he added, was to appeal to the power of God. Perhaps Riccioli had Digges or Lansbergen in mind. That's a bogus argument, he said, one that, while it cannot be refuted (who can deny the power of God?), "cannot satisfy the more prudent sort of person."[21]

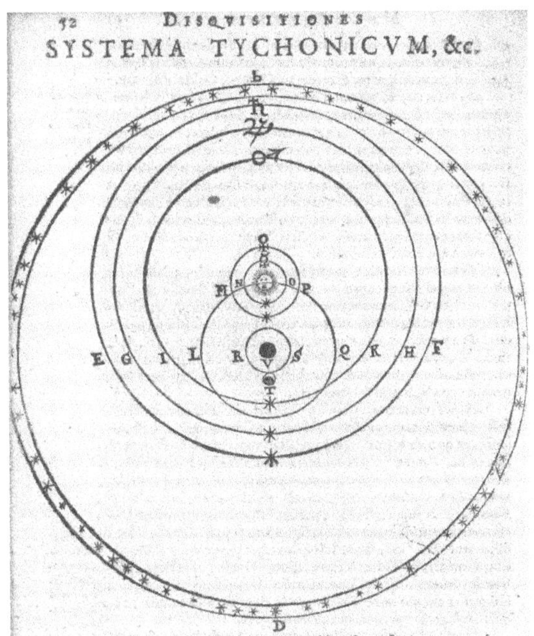

Figure 8.7 Illustration by Locher and Scheiner of the geocentric model of the Universe developed by Tycho Brahe. Sun, Moon, and stars circle an immobile Earth. Stars are located just beyond Saturn. Planets circle the Sun. This model, also called a *geo-heliocentric* model, could accommodate all of Galileo's telescopic discoveries, could retain key ideas from Aristotle and Ptolemy, and was free from problems of parallax and stellar distances and sizes and of any Coriolis Effect.
Credit: ETH-Bibliothek Zürich.

[21] Riccioli loosely translated here. See Graney, *Setting Aside*, 133, 137.

Anti-Copernicans had recourse to an alternative geocentric model of the Universe, developed by Tycho Brahe, in which the Sun, Moon, and stars orbit a fixed Earth, but the planets circle the Sun (see Figure 8.7). It essentially preserved the two-storey Universe. It was entirely compatible with telescopic discoveries, and yet it avoided all the Copernican problems involving physics, deflecting projectiles, and parallax/giant stars.

Copernicans did not have to turn to God to answer the anti-Copernicans; there was always the option of bluster and obfuscation. In his *Dialogue*, Galileo spent pages and pages on stars and their sizes. Salviati, his Copernican mouthpiece, said a great deal. But the expenditure of words didn't work because you just can't argue with reproducible data. Tacquet would later charge that "Galileo in his *Dialogue* attempts in vain to dodge the monstrous sizes of the fixed stars. He gives a lengthy discourse—thirty-three pages—but it does nothing to undermine our mathematically-based demonstrations."[22]

Tacquet was right. Really, all Copernicans could do was embrace the giant stars and assert that these vast bodies declare the power of God! That in effect is also what Kepler did, logically concluding that in a Copernican Universe the stars therefore could absolutely not be other Suns. And he was very, very happy about this.[23]

---

[22] Tacquet loosely translated here. For a full scholarly treatment of Taquet and of Galileo's arguments in the *Dialogue*, see Graney, "Galileo Between Jesuits," where Taquet is more precisely quoted on 201–202.

[23] Interested readers may compare this chapter with Dennis Danielson and Christopher M. Graney, "The Case Against Copernicus," *Scientific American* (January 2014): 72–77; our knowledge of the "case" has grown considerably in the intervening decade.

9

Dethroning the Sun

When Johannes Kepler first heard about the contents of Galileo's *Starry Messenger*, he was *not* happy. He got the word from a friend who had stopped by to gloat. It was mid-March of 1610, and Johannes Matthaeus Wackher von Wackenfels, counselor to the Holy Roman Emperor Rudolph II, arrived at Kepler's house with news of "four previously unknown planets, discovered by the use of the telescope with two lenses," as Kepler would write in an open letter to Galileo that April.

But if Wackher was "overcome with joy by the news," Kepler was overcome "with shame." It looked as if he was about to lose an argument.

Wackher and Kepler, both Copernicans, had been arguing about the existence of other planets. Kepler believed that the Solar System was unique, with the spacing of its six planets (Mercury, Venus, Earth, Mars, Jupiter, and Saturn) set by God through an arrangement of the five Platonic solids (see Figure 9.1). Wackher, on the other hand, maintained that other planets "undoubtedly" circle some of the fixed stars. He had been reading the "speculations" (as Kepler put it) of Giordano Bruno. Given this news of four new planets up there, Wackher was saying, why should we not expect to find countless others in the future? The Universe is infinite, and there is an infinite number of other worlds—or, "as Bruno puts it," other earths.[1]

But when Kepler finally read *The Starry Messenger*, he found out that the "four planets" Galileo had discovered were not circling a fixed star. They were circling Jupiter. Kepler wrote to Galileo,

> I rejoice that I am to some extent restored to life by your work. If you had discovered any planets revolving around one of the fixed stars, there would now be waiting for me chains and a prison [prepared by the gloating Wackher] amid Bruno's innumerabilities, I should rather say, exile to his infinite

---

[1] *KCGSM* [Kepler, *Kepler's Conversation with Galileo's Sidereal Messenger*, tr. Rosen], 10–11.

A Universe of Earths. Dennis Danielson and Christopher M. Graney, Oxford University Press.
© Oxford University Press (2025). DOI: 10.1093/9780197803547.003.0009

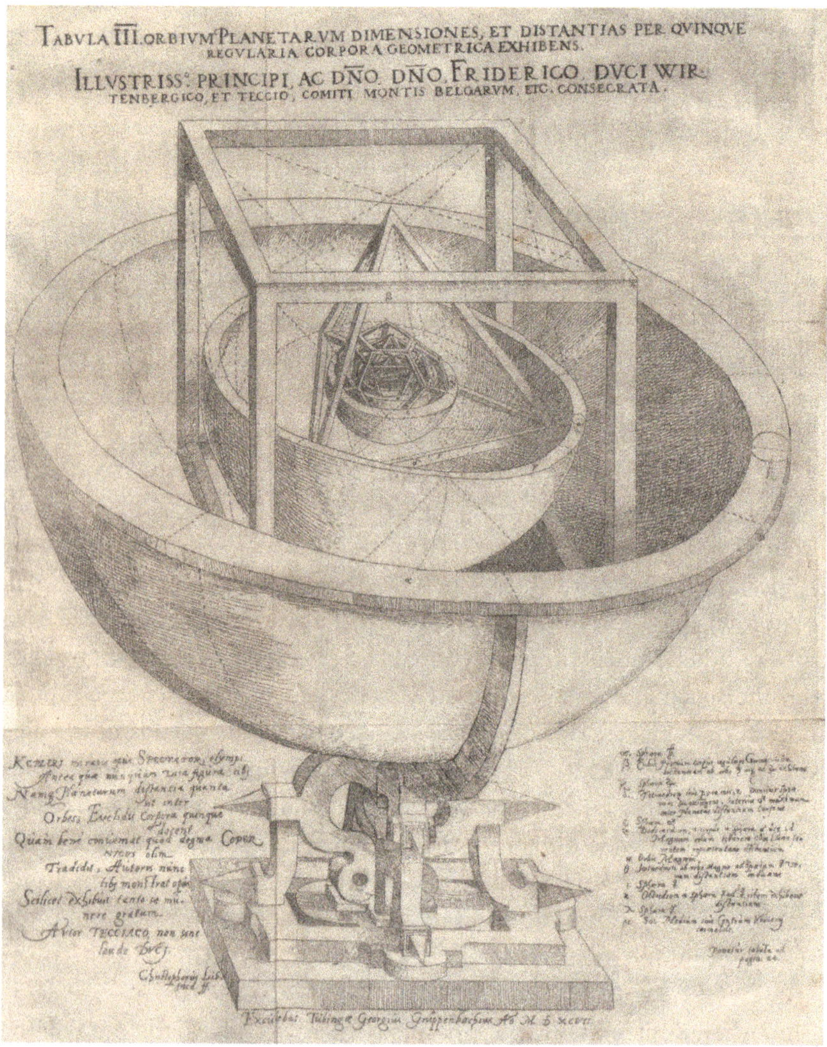

Figure 9.1 Kepler's "Cosmographic Mystery" showing that God had spaced the planets according to a nesting of the five Platonic solids (cube, tetrahedron, etc.).

space. Therefore, by reporting that these four planets revolve, not around one of the fixed stars, but around the planet Jupiter, you have for the present freed me from the great fear which gripped me as soon as I had heard about your book from my opponent's [Wackher's] triumphal shout.[2]

2 *KCGSM*, 36–37.

Nothing has yet been seen revolving around the fixed stars, Kepler says. "Hence this will remain an open question until this phenomenon too is detected by someone equipped for marvellously refined observations."[3]

Kepler was also happy to read in the *Starry Messenger* about something else: Galileo's observations of the fixed stars. These were "highly welcome." First, the sparkling brilliance of the stars compared to the duller planets (when seen telescopically) shows that they generate their light "from within," whereas the planets are illuminated "from without." In *Bruno's* terms, Kepler says, stars are "suns," planets are "moons or earths."

But in fact, Kepler can demonstrate that stars are not really suns at all. Let Bruno not "lead us on to his belief in infinite worlds, as numerous as the fixed stars and all similar to our own." Another of Galileo's star observations matters here. "You do not hesitate to declare that there are visible over 10,000 stars," Kepler writes. "The more there are, and the more crowded they are, the stronger becomes my argument against the infinity of the universe."

Now Kepler really pours it on: "Suppose that we took only 1,000 fixed stars, none of them larger than 1 minute of arc [1/30 the diameter of the moon]." The majority of stars in the catalogues are larger than this, he says.

> If these were all merged in a single round surface, they would equal (and even surpass) the diameter of the sun. If the little disks of 10,000 stars are fused into one, how much more will their visible size exceed the apparent disk of the sun? If this is true, and if they are suns having the same nature as our sun, why do not these suns collectively outdistance our sun in brilliance?

In other words, contrary to Bruno, the stars clearly are not suns. They do not compare to the Sun in light output. Even merely a thousand stars together would form a disc in the sky with an apparent size comparable to the Sun, and yet their light is as nothing compared to it: "When sunlight bursts into a sealed room through a hole made with a tiny pin point, it outshines the fixed stars at once. The difference is practically infinite." And Galileo has shown that there are far more stars than previously numbered.[4]

The stars are not only too dim, but also too big, to be suns. In his 1604 book *On the New Star*, Kepler had calculated that Sirius, the Dog Star, the brightest star in the sky, must be bigger than the orbit of Saturn—bigger than the

<hr>

[3] *KCGSM*, 39.
[4] *KCGSM*, 34–35.

entire Earth-centered Universe of Tycho Brahe (see Figure 8.7). And if Sirius were that big, every visible star in the sky would be bigger than the orbit of Earth. Kepler was a Copernican who embraced giant stars. The giant stars help us recognize "the immensity of divine power," he had said in *New Star*. At the same time, such is the beauty and vibrancy of Earth with its "fine bits of dust" called human beings (recall from Chapter 1), who have received such gifts from the Creator that we can see also how God "ennobles those whom He willed to be small."[5]

Might Wackher claim that the dimness of the stars is caused by their distance? "This does not help his cause at all," Kepler says to Galileo, "For the greater their distance, the more does every single one of them outstrip the sun in diameter." In *On the New Star*, he writes: "imagine a star which seems to subtend a definite angle, say, 4 minutes [4/30 the diameter of the Moon]. The width of such a body is always one-thousandth of its distance, as is absolutely certain from geometry."[6] So, if the star is a million miles distant, its diameter is a thousand miles—if it is a billion miles distant, its diameter is a million miles. The farther, the bigger. And distance would have no effect on how bright the star would appear. The surface brightness of an object does not decrease with distance; when you stand in a grassy field on a sunny day, a square foot of grass at your feet is not brighter than a square foot twenty yards off; the ground does not grow dimmer with distance. A glowing lampshade, illuminated by a bulb within, does not darken as you move away from it. Kepler knew what he was talking about.

Thus, Kepler used simple observations, measurements, and calculations—the most basic *science*—to demonstrate that stars are dim but giant bodies. In a Copernican Universe, every visible star could hold within itself at least the Sun, with Mercury, Venus, and Earth all orbiting it; Sirius could hold the whole Solar System. But the Sun, a tiny body compared to even the smaller stars, outshines them all combined, by a "practically infinite" amount. What to Galileo's Sagredo would be "entirely unbelievable" Copernican stars, and what to Brahe was an absurdity of the Copernican system, was to Kepler a ticket out of Wackher's prison. The Solar System was unique.

Everyone with clear eyes and basic knowledge of measurement and geometry could do the science for themselves and come to the same conclusion. There really was no arguing with it. There was absolutely no science involved

---

[5] Kepler, Chapter 16 of *De Stella Nova*, 58.
[6] Kepler, Chapter 21 of *De Stella Nova*, 129.

in Wackher's support for Bruno's ideas—Wackher just found them appealing, much as Horrocks or von Guericke found appealing the idea of planet size increasing regularly with distance from the Sun (recall from Chapter 1), despite the lack of scientific evidence for it. Yet science or no, Bruno's ideas about stars being suns would appeal to more people than just Wackher.

René Descartes (1596–1650), a younger contemporary of Kepler, helped these ideas by devising a mechanistic explanation of *why* orbits work (in contrast to Kepler, who developed the mathematics of *how* they work). Descartes proposed that the Universe is filled with a sort of fluid matter in which all planets, moons, and so forth are immersed, such that orbital motion results from whirlpools or vortices in this fluid.[7]

Accordingly, planets orbit a sun for the same reason that soap bubbles in water swirling down a fast drain orbit the drain—they are caught in a vortex. Our Sun sits in the center of a large vortex that carries its planets around it. Jupiter with its four moons and Earth with its one Moon sit within smaller vortices that exist within the solar vortex. Other suns would have their own vortices.

Several decades after Kepler and Descartes, Bernard Le Bovier de Fontenelle borrowed the vortex idea for his 1686 *Conversations on the Plurality of Worlds*, which built on and promoted the idea of a Universe of other suns and other earths. It featured a frontispiece by Juan d'Olivar, showing the Solar System as but one system among a vast swarm of suns and planets. "French society was shocked," says art historian Lucía Ayala, and "this shock quickly spread throughout Europe"[8] (see Figure 9.2).

Conversations, as you may recall, was written in the form of a discussion between a narrator and a Countess. Among other apparently surprising ideas, they discuss how the Moon must be inhabited because it is like the Earth, and the planets as well, because they are like the Moon. The stars—being "luminous bodies in themselves, and so many suns," according to the narrator—would be centers of their own myriad vortices, with their own planets featuring inhabitants looking out upon the Universe just as we earthlings do. The Universe is filled, as d'Olivar's illustration shows, with many suns and many earths—filled, that is, with a "Plurality of Worlds" like ours.[9]

---

[7] Charalampous, "'One common matter' in Descartes' physics."
[8] Ayala, "On the Plurality of Worlds."
[9] Fontenelle, *Conversations*, 105–106.

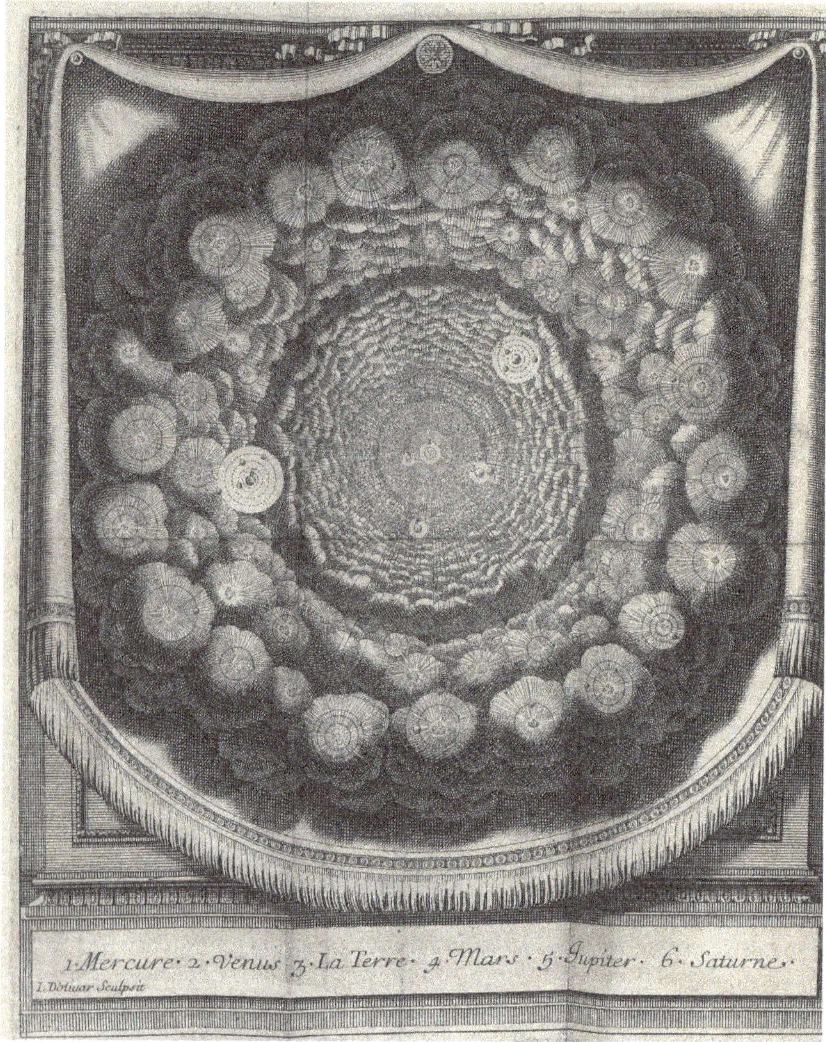

Figure 9.2 Fontenelle's frontispiece.

Credit: The Linda Hall Library of Science, Engineering & Technology

No doubt Kepler spun in his grave over Fontenelle's easy assumption that the stars are "so many suns." Fontenelle's reasoning was none too scientific: because "their light is bright and shining," he said.[10] He was not writing a work of rigorous science. His narrator claims, for example, that the

[10] Fontenelle, *Conversations*, 104.

interactions between the vortices of the stars are the cause of their twinkling, apparently unaware that an interested observer can easily ascertain that the twinkling of stars is a result of Earth's own atmosphere (twinkling is obviously influenced by the weather), and furthermore that this phenomenon was fully addressed seven decades earlier by Locher and Scheiner in their *Disquisitions.*[11]

Conversations may have lacked scientific rigor, but Ayala makes the case that it eventually had a certain political resonance. King Louis XIV of France (who reigned 1643–1715) had embraced Copernicanism as a metaphor for his rule as "The Sun King." Recall the Copernican idea of the Sun, which (as Digges worded it) "like a king in the midst of all reigneth and giveth laws of motion to the rest, spherically dispersing his glorious beams of light through all this sacred celestial temple." Ayala shows how Louis XIV endeavored "to convince others of his own conviction: that natural forces legitimized his power, Copernicanism being his major proof"[12]—as illustrated by the flattery that Pierre Le Lorrain offered King Louis in the preface to a 1707 book about astronomical systems:

> Indeed, SIRE, you are at the centre of this kingdom as the sun, according to the hypothesis of Copernicus, is at the center of the universe. The respectable serenity, which is always seen glowing on your noble face, in situations when Philosophy would be disconcerted, shows us how religion has elevated your sentiments above nature; and to us you are a vivid image of this perfect repose, as Copernicus represents the sun inscribed from the centre of the world out to its circumference, whose powerful virtue makes the Earth and all the celestial globes move. Is your state not ordered like the visible world?[13]

How thoroughly renovated was the old cosmic basement! The sump was now the throne room. But the huge irony is that Fontenelle's 1686 work, although published *avec privilege du Roy*, eventually undermined the "single Sun" Universe that the Sun King found so politically useful as a metonym of his reign. In Ayala's words, he "underestimated the . . . unstoppable revolutionary potential of the new complex universe [Fontenelle]

---

[11] Fontenelle, *Conversations*, 115–16; Locher and Scheiner, "Disquisition 34," *MathDisq* [Locher with Scheiner, *Disquisitiones Mathematicae*, tr. Graney], 78–83.

[12] Ayala, "Cosmology after Copernicus," 207.

[13] Ayala, "Cosmology after Copernicus," 209.

describes."[14] For if the stars are all other suns, then the government exercised by our Sun becomes a much diminished and merely parochial affair. D'Olivar's illustration features a face in every one of the innumerable suns (see Figure 9.3). Fontenelle's book thus implies that our Sun is not special and that the Universe is *not* structured around a single, brilliant ruler.[15]

Fontenelle's was an achievement much more of imagination than of science. Yet by the time *Conversations* was published, science had ceased to be as solidly against the idea of the stars as other suns as it was in Kepler's time. Kepler's single-Sun Universe, as we've seen, hinged on the apparent sizes of stars. And while he loved the idea of giant stars, the fact was that giant stars were a giant problem for the supporters of Copernicus. Anti-Copernicans saw them as an absurdity, something hatched up to wave away the fact that no parallax was visible in the stars. The solution to the parallax problem was not distant, monster stars, anti-Copernicans said; the solution to the parallax problem was to acknowledge that Earth did not circle the Sun and that Tycho Brahe's model of the Universe (see Figure 8.7) was the true one.

But by the 1660s, new lines of evidence suggested that human eyes and even telescopes radically inflated the apparent sizes of stars. And if this were truly the case, then stars *could* be radically smaller[16]—they could be suns after all.

This evidence did not go unchallenged. For example, when Dutch astronomer Christiaan Huygens published observations suggesting that

Figure 9.3 Fontenelle's frontispiece (detail).

[14] Ayala, "Cosmology after Copernicus," 213.
[15] Ayala, "Cosmology after Copernicus," 216.
[16] Graney, *Setting Aside*, 148–54; Graney, "Galileo between Jesuits," 220–24.

telescopes inflated the sizes of stars, John Flamsteed, England's first Astronomer Royal, countered that Sirius and Mercury, seen through the same small-aperture telescope under the same conditions, had identical apparent diameters. That would make Sirius as much larger than Mercury as it is farther away—and pretty much any Copernican distance estimate for Sirius was enough to make Sirius gigantic.[17] Thus, when Fontenelle published *Conversations* in 1686, the question of whether science was against the idea of stars as suns was very much in flux.

Regardless of the science, however, enthusiasm for a Plurality of Worlds, or "Pluralism," had caught on. In 1716 and again in 1740, Jacques Cassini of the Paris Observatory published high-quality telescopic measurements of Sirius that showed it to be giant, granted a Copernican Universe. Just like Kepler, he showed that all stars had to be giant compared to the Sun—that they were therefore not other suns.[18]

It mattered not. Although fellow astronomers acknowledged Cassini's measurements, they nevertheless continued to speak of the stars as other suns. For example, John Hill acknowledged Cassini's observations in his 1754 encyclopedic *Urania*, writing that

[the] observation of Sirius's diameter . . . had, for its author, one of the most accurate, and most judicious astronomers the world has ever known, Cassini, and, whenever it is repeated with the same apparatus, it succeeds in the same manner, and verified very punctually; and other stars have also apparent diameters of nearly the same extent.[19]

Note the emphasis on the reproducibility of Cassini's results. Nevertheless, Hill persisted in describing stars as other suns:

The more we see of the works of the creation, the more we must admire and adore its Author. It does appear that the unbounded space is filled at proper distances with these stars: each of these is a sun; and if we continue the inquiry, reasoning by analogy, we shall determine that each of these suns has earthy planets rolling round it, for to what end else should they have been created? In this view, what, and how amazing is the structure of the universe![20]

[17] Graney, *Setting Aside*, 148–58.
[18] Graney, "The Starry Universe of Jacques Cassini"; Graney, "Galileo between Jesuits," 223–24.
[19] "STARS, fixed" in Hill, *Urania*, fol. Z2ᵛ.
[20] "DISTANCE of the fixed stars," Hill, *Urania*, fol. Z2ʳ.

In addition to the growing political appeal of a Plurality of Worlds in the aftermath of the Sun King's reign, Hill's words suggest that perhaps there was also religious appeal. And when he describes the Plurality of Worlds as an "amazing" work of creation, a tribute to its "Author," he is clearly not envisioning a Universe of "earthy" sumps or even "earthy" little dark stars. No, the Plurality that Hill is envisioning is something akin to Galileo's "dance of the stars," a dynamic cosmic community of *admirable* "earthy" worlds. Whatever the reason, Hill would not have Pluralism be deflated by a little high-quality, easily reproducible telescopic data from a highly skilled scientist.

Kepler, with science behind him, had scoffed at the idea of a Plurality of Worlds, with myriad "Galileos, observing new stars in new worlds."[21] But Fontenelle advocated for that Plurality and its many stargazers looking out upon the Universe. And science seems to have been no match for the appealing prospect of envisioning amazing worlds and dismantling cosmic (and earthly) hierarchies.[22]

[21] *KCGSM*, 44.
[22] For a reliable and accessible introductory survey of the history of Plurality, see Dick, *Space, Time, and Aliens*, Chapter 1.

10

"Reasoning by Analogy"

Earths, Suns, Galaxies

"Human beings are analogy machines . . . dealing with new information by comparing it to things they already know something about."[1]

Our human search for knowledge regularly proceeds by means of analogies, similes, and likenesses. A friend experiences a new place or meets a new person, and you ask her: "What was it *like*?" or "What was he *like*?" If the place or person is utterly unique, then the question simply can't be answered. But we persist in seeking some kind of parallel, some degree of analogy. And in science as well as everyday experience, once we do discern similarities between things or phenomena, we use these as a kind of beachhead from which to draw further inferences about possibly shared features or properties.

This manner of inference plays a huge role in the history of astronomy and cosmology, and one of its most notable manifestations is the way that things once considered unique acquire plural forms. Right from his midtwenties, John Wilkins, whom we met in Chapter 5, discerned the exciting analogy offered by the Copernican picture of the world, one that continues to resonate today: "if our earth were one of the planets," he asked, "then why may not another of the planets be an earth?"[2]

The telescope had shown that, contrary to Aristotle's physics, there was another center in the Universe, not just *the* center of the Universe, Earth's location. There were at least two others—the changing phases of Venus showed its orbit to be centered on the Sun, and Jupiter was clearly the center of its own system. That Jovian system also meant that there was another moon, not just *the* Moon. Four moons circled Jupiter.

---

[1] See OED, "analogy," II.7.b. The epigraph is from the *New York Times*, March 23, 2008. Business Section 4/1.

[2] Wilkins, *The Discovery of a World in the Moon*, 93–94.

A Universe of Earths. Dennis Danielson and Christopher M. Graney, Oxford University Press.
© Oxford University Press (2025). DOI: 10.1093/9780197803547.003.0010

Humans could now speak of *centers* and *moons*. Other astronomical nouns similarly took on plural forms, but through analogy, not through telescopic observation. As we have seen, in the late 1500s, the anti-Aristotelian Giordano Bruno extrapolated (contrary to Copernicus, and, as Kepler emphasized, contrary to observational evidence) an infinite Universe containing "innumerable suns, and likewise an infinite number of earths circling about those suns."[3] And by the middle of the eighteenth century John Hill, "reasoning by analogy," was producing astronomical conclusions about widely distributed "suns," each with "earthy planets rolling round it"—conclusions that *preceded* any actual telescopic evidence of other planets around other suns (which was Kepler's nightmare when in 1610 he heard about Galileo's discovery of "planets"). Nor would the process of analogy end with only Moon, Earth, and Sun acquiring plurals.

Recognizing analogy's tendency to get ahead of empirical science is not to denigrate its power but to suggest caution regarding its application. For, as the logician L. S. Stebbing puts it, "We are all at times apt to be impressed by weak analogies owing to the fact that a resemblance that is *interesting* or emotionally satisfying will seem to us to be significant if we are not on our guard. The greater our ignorance of the subject-matter, the more likely we are to be misled by a weak analogy."[4] And yet, even if analogies fail to constitute scientific or philosophical proof, they can act as a mighty stimulus to the exploration of many things, including the Universe.

This was certainly true of the great eighteenth-century German philosopher Immanuel Kant (1724–1804), whose *Universal Natural History and Theory of the Heavens* appeared in 1755, mere months after publication of Hill's comments in *Urania* about "reasoning by analogy." Kant had read a lengthy German review of the Englishman Thomas Wright's *An Original Theory or New Hypothesis of the Universe* (1750), which adumbrated a theory of the Milky Way by way of analogy with the structure of the Copernican Solar System.

The Milky Way, that band of faint light that bisects the heavens, is not very familiar to the average person today, thanks to waste lighting at night pouring upward into the sky and washing it out. But under the sorts of truly dark skies that were once common even in cities, it is plainly visible and

---

[3] Bruno, *De l'infinito universo*; translation from *BOTC* [*The Book of the Cosmos: Imagining the Universe from Heraclitus to Hawking*, ed. Danielson], 144.

[4] Stebbing, *A Modern Introduction to Logic*, 254. See also Bartha, "Analogy and Analogical Reasoning."

is quite dramatic. The Greek word for "milky" is "*galakticos*," and thus that band was called *the* Galaxy.

Recall from Chapter 4 that one of the things that Galileo discovered with his telescope was that the Galaxy was not, in fact, a continuous band of light. It was made up of vast numbers of stars. "Point your telescope in any direction within the Galaxy," he said, "and at once a great mass of stars comes into view."[5]

These vast numbers of stars, Wright proposed, move about a common center just as the known planets move about the Sun, for which reason "we consider [the stars] all as flaming suns, progenitors, and *Primum Mobiles* of a still much greater number of peopled worlds."[6] The brilliant thirty-year-old Kant picked up Wright's unsystematic account as reported in a German book review. And he wrote his own book, unfolding a cosmology still grander than Wright's by means of three stunning analogies.

The first of these analogies develops Wright's parallel between the structure of the Solar System and the superstructure of the Galaxy. Kant appeals to Newton's laws of gravitation such that

> every world-system is constituted by a mutual drawing together, which persists unceasingly and utterly unhindered, and whereby sooner or later each implodes into a single clump, unless this cataclysm is prevented, as it is with the spheres of our planetary system, by the action of centrifugal forces. These forces, by deflecting the heavenly spheres from falling in a straight line, produce, in combination with those forces of attraction, the eternal orbital revolutions whereby the edifice of the creation is secured from destruction.[7]

Note: "implodes into a single clump." According to Aristotle, gravity was a tendency of earth and water to move toward a specific point, *the* center of the Universe. When you drop a rock, it falls downward to the ground. For Aristotle, the ground, the Earth, is merely the accumulation of all the stuff that tends toward the center. Take it all away, and the rock would still fall—toward a point, not toward the Earth as such. If you took the rock to the surface of the Moon and let it go, it would still fall toward that center

---

[5] *BOTC*, 152.

[6] Wright's book is excerpted in *BOTC*, 259–64.

[7] Quotations of Kant's *Universal Natural History* parts 1 and 2 are drawn from *BOTC*, 265–71, where the translation is adapted from Kant, *Kant's Cosmogony*.

point, away from the Moon (which, being of the ethereal upper storey of the Universe, does not tend toward the center).

By contrast, Isaac Newton proposed that the rock falls toward the ground because gravity is a universal attraction between mass-having objects, an attraction that decreases with greater distance and lesser mass. The rock thus falls toward the Earth, not toward a point. Take the Earth away, and the rock would not fall. Take the rock to the surface of the Moon, and it would fall to the Moon, because the gravitational attraction of the much-closer Moon would be greater than that of the distant-but-larger Earth. Even before Newton, Kepler had argued in 1611 that there was gravity on the Moon, as evidenced by the "very great lakes" seen there with the telescope: "the quantity of water on the moon is not impeded at all from adhering steadfastly to the sphere of the moon—that is, from directing itself toward the center of its body."[8] Kepler was wrong about the Moon having lakes of water, but he was right in his intuition that it does have gravity.

The Moon itself is gravitationally attracted to Earth. It does not fall because it is in motion: Newton's (and Kant's) "centrifugal forces" keep it up there. Stop the Moon in its orbit, however, and it will fall to Earth; the Earth–Moon system will "implode"—likewise for the Jovian system and for the Solar System as a whole.

This is a significantly less stable picture than that of the two-storey Universe, in which the ethereal Moon stays up and moves, by its very nature: In the two-storey Universe, the Moon cannot conceivably fall; in a Newtonian world, unless forces are finely balanced, it just might.

Thus, Kant concludes that the Solar System formed from the gravitational in-fall of a diffuse mass, "as raw and undeveloped as possible," mitigated only by the centrifugal effects of the resulting motion. The Milky Way, according to Kant, formed in this way too: "The shape of the heaven of fixed stars thus has no cause other than the same systematic arrangement on a grand scale as the cosmic structure of the planetary system on a small scale." It developed into a kind of dynamic disk of stars circling in roughly the same plane. This is why we on Earth see the Galaxy as a band of light. We are in the midst of this disk comprising vast numbers of stars. We see it "edgewise," appearing (according to the analogy) as a band of innumerable self-luminous bodies, "on account of their proximity to the common configuration within which they are related—we on earth also being located in that same plane,"

---

[8] Kepler, "Dioptrics," 204.

which forms a "densely lit area" that we see as encircling us. "This band of light would be seen everywhere copiously set about with stars, although according to this hypothesis they are wandering stars and *not fixed* in one place."

The italics for the words *not fixed* are added. Kant's hypothesis actually destroys the very notion of "fixed stars." They are not fixed at all, but rather are in dynamic motion. Ancient and medieval teachings about the unchanging perfection of the Universe beyond the sphere of the Moon, already undermined by Tycho's "new star" and by telescopic observations of the Sun and Moon, must now be discarded on a cosmic scale and replaced by possibilities of inherent cosmic change and cosmic evolution. Kant admits that the stars' designation as "fixed" has been "unshaken by centuries of observation." Yet, he explains, this mere appearance arises either from the stars' "extraordinary slowness" or else from the "mere imperceptibility [of motion] occasioned by the distance of that place from which the observation is being made."

Given his own and his readers' sheer familiarity with the term, however, Kant continues to refer to "fixed" stars as he moves to his next, breathtakingly prescient application of analogy. He tackles the nature and location of nebulous stars, which have come to be referred to simply as *nebulae*. These "cloudy" stars (*nebula* is the Latin word for "little cloud") can be seen throughout the sky (if the sky is dark). Two very prominent examples, both visible to the naked eye, are found in the sword of the constellation Orion and in the constellation Andromeda; the Andromeda nebula is wider than the Moon. Most nebulae are much smaller than Andromeda, however, and are visible only telescopically.

"These nebulous stars," Kant writes, "must undoubtedly be removed at least as far from us as the other fixed stars, [so that] not only would their magnitude be astonishing . . . but also it would be most peculiar if, despite their extraordinary size, such self-luminous bodies and suns were to give off the dullest and feeblest light." Instead, Kant proposes, they should be considered "not such enormous single stars but rather systems of many stars, whose distance presents them within such a narrow space that their light, which from each one individually is imperceptible, strikes us on account of their immense multitude as a pale glimmer." Therefore, Kant asserts, "these elliptical [as seen through a telescope] figures are just such world-systems and, so to speak, Milky Ways, whose structure we have just unfolded."

World-systems, milky ways, other galaxies: Kant's account of the nebulae deserves huge credit for prescience, especially when we consider that the question of whether other Milky Ways, other galaxies, actually exist was not settled scientifically until the twentieth century. Kant's mid-eighteenth-century proposal that the nebulae are in fact other galaxies was truly monumental. Analogy alone did not guarantee its correctness. The Andromeda nebula truly is another galaxy (its distance, measured by Edwin Hubble in the 1920s building upon the work of Henrietta Leavitt, proved it is not in our Milky Way). The Orion nebula is not another galaxy but is truly a cloud of dust and gas within our Galaxy. Yet the proposal itself well illustrates the powers of analogy.

Kant's third exercise of analogy is less well known than his proposal about the nebulae. The third part of his *Universal Natural History* is subtitled "an attempt, based on natural analogies, to establish a comparison between the inhabitants of different planets."[9]

One of Kant's predecessors in making such an attempt was the Dutch astronomer Christiaan Huygens, whose posthumous *Cosmotheoros*, or *Celestial Worlds Discover'd* (1698) analogized and speculated concerning "planetarians." Huygens wrote that "A man that is of Copernicus's opinion, that this earth of ours is a planet carried round and enlightened by the sun like the rest of them, cannot but sometimes have a fancy that it's not improbable that the rest of the planets have their dress and furniture, nay and their inhabitants too as well as this earth of ours."[10] Huygens justifies the use of analogy by means of another analogy:

If anyone should be at the dissection of a dog, and be there shown the entrails, the heart, stomach, liver, lungs and guts, all the veins, arteries and nerves; could such a man reasonably doubt whether there were the same contexture and variety of parts in a bullock, hog, or any other beast, though he had never chanced to see the like opening of them? I don't believe he would. . . . 'Tis therefore an argument of no small weight that is fetched from relation and likeness; and to reason from what we see and are sure of, to what we cannot, is no false logic.[11]

---

[9] In this part of our discussion of Kant, we use the translation by Ian Johnston, https://web.viu.ca/johnstoi/kant/kant2e.htm. Thanks to the translator for providing us with a file containing this version.

[10] Huygens, *Celestial Worlds Discover'd*, 1–2.

[11] Huygens, *Celestial Worlds Discover'd*, 18.

Kant's pursuit of the same theme of planetarians is marked by his legendary powers of analysis. His discussion exudes some of the kinds of argument we still hear today—and will return to in subsequent chapters—as they concern ideas about extraterrestrial life and intelligence.

One of Kant's insights is that whatever beings inhabit other worlds must be adapted to the conditions of those worlds. Such beings' "way of working and suffering is associated with the composition of the material to which they are bound and depends upon the quantity of impressions which the world arouses in them, according to the relationship of their living environment with the centre of the power of attraction and heat." Moreover, having recognized change and development throughout the Cosmos ("the drama of the passing changes of the Universe"),[12] Kant postulates that worlds might pass through stages during which they are not yet capable of supporting life. Although we may smile at the ludicrously low numbers he employs—in contrast to today's understanding of vast epochs of geological change—he nonetheless astutely observes that "[o]ur Earth perhaps existed for a thousand years or more before it was in a condition to be able to support human beings, animals, and plants."[13] And Jupiter, he avers, is apparently still a "long way from having a calm outer surface, a condition which must pertain for a planet to be inhabited."

Granted, however, that we inhabitants of Earth and the constitution of our planet are admirably adapted to each other, Kant enlists another analogy—this time a humorous one—to prevent too prideful a human estimation of our place in the Universe. He cites a writer whom he identifies only as a "witty person from the Hague," who offers this scenario:

> Those creatures who live in the forests of a beggar's head . . . had for a long time thought of their dwelling place as an immeasurably large ball and themselves as the masterworks of creation. Then one of them, whom Heaven had endowed with a more refined soul, a small Fontenelle of his species, unexpectedly learned about a nobleman's head. Immediately he called all the witty creatures of his district together and told them with delight: "We are not the only living beings in all nature. Look here at this new land. More lice live here."

[12] Kant: "das Schauspiel der ablaufenden Veränderungen des Universi."
[13] Ian Johnston translation.

"This insect," Kant reflects, "which in its way of living as well as in its lack of worth expresses very well the condition of most human beings, can be used for such a comparison with good results. Since, according to the louse's imagination, nature is endlessly well suited to its existence, it considers irrelevant all the rest of creation which does not have a precise goal related to its species as the central point of nature's purposes."

The reference to Fontenelle, whose *Conversation on the Plurality of Worlds* we considered in Chapter 7, hints at an increasing capacity of astronomy, with its declarations of ever greater cosmic distances and magnitudes, to put humankind in its place—ethically as well as cosmologically. Referring to Copernicus, Fontenelle wrote: "I am extremely pleased with him . . . for having humbled the vanity of mankind, who had usurped the first and best situation in the universe. And I am glad to see the earth under the same circumstances with the other planets." Here we discern a misanthropic undercurrent in discussions of other worlds, echoed too in Huygens's rhetorical query regarding "whether Nature has laid out all her cost and finery upon this small speck of dirt." These sentiments are in stark contrast to Kepler's "fine bits of dust" view of humankind.

Kant may take the astronomically-based opportunity for disapproval of humankind even further than his predecessors when he compares our species unfavorably with the lice who "inhabited forests on the beggar's head." Did these lice ever create "greater disasters among the races of this colony than the son of Philip [Alexander the Great] brought about among the race of his fellow citizens, when his wicked genius gave him the idea that the world was created only for his sake?" As we'll see, what we're calling astronomy's misanthropic undercurrent is still part of the popular scientific landscape.

Kant's increasingly speculative discourse in the *Universal Natural History* culminates in his assertion—in keeping with his vision of a changing, we might say *evolutionary* Cosmos—that "most of the planets are certainly inhabited, and those that are not will be in the future." Furthermore, Kant claims, "the material stuff out of which the inhabitants of different planets, including even the animals and plants, are made must, in general, be of a lighter and finer type, and the elasticity of the fibres as well as the advantageous construction of their design must be more perfect in proportion to their distance away from the sun" (Echoes of Horrocks and von Guericke!)

Among (in increasing distance) Mercury, Venus, Earth, Mars, Jupiter, and Saturn, we on Earth accordingly find ourselves in something of a mediocre position. For us humans, however, Kant seems to consider that this picture offers some consolation: "If the idea of the most sublime classes of sensible creatures living on Jupiter or Saturn provokes the jealousy of human beings and discourages them with the knowledge of their own humble position, a glance at the lower stages brings content and calms them again. The beings on the planets Venus and Mercury are reduced far below the perfection of human nature."

Finally, Kant offers still further consolation: an almost apocalyptic vision in which we on Earth might someday rise above our terrestrial position:

Is the everlasting soul for the full eternity of its future existence . . . always to remain fixed at this point of the cosmos, on our Earth? Is it never to share a closer look at the rest of creation's miracles? Who knows whether it is not determined that in future the soul will get to know at close quarters those distant spheres of the cosmic structure and the excellence of their dwelling places . . . ? Perhaps that is why some spheres of the planetary system are already developing, in order to prepare for us in other heavens new places to live after the completion of the time prescribed for our stay here on Earth. Who knows whether those satellites do not circle around Jupiter so as to provide light for us in the future?

Despite its highly speculative character, Kant's work offers a powerful set of fresh responses to a Universe whose dimensions and potentialities were in the eighteenth century continuing to expand as the implications of the Earth being part of the "dance of the stars" sank in—as they *continue* to expand. And these responses at the same time clearly illustrate the power as well as the possible limitations of analogical reasoning. As with the examples of Kant's claims about other planets' inhabitants, we are left (to echo Stebbing) with "*interesting* or emotionally satisfying" claims that, upon examination, conspicuously lack scientific evidence.

11

The Problem of Life

Immanuel Kant hypothesized that the Solar System and the Galaxy were the result of Newton's laws of gravitation and motion, that "mutual drawing together, which persists unceasingly and utterly unhindered," but in fact *is* hindered "by the action of centrifugal forces." This idea had some interesting consequences for the Plurality of Worlds regarding life. Sir William Thomson (later Lord Kelvin) talked about some of these in an address he gave at the forty-first meeting of the British Association for the Advancement of Science (BAAS), held in Edinburgh, Scotland, in August of 1871.

Thomson was the president of the association at the time. He gave a long address—roughly 15,000 words (the combined length of several chapters of this book)—about the BAAS, the prominent scientists of the past who had been part of it, and the progress of science since its formation. He devoted the last quarter of the talk, however, to the subject of origins. "The old nebular hypothesis," he said, speaking of Kant's idea, "supposes the solar system and other similar systems through the universe which we see at a distance as stars, to have originated in the condensation of . . . nebulous matter. This hypothesis was invented before the discovery of thermo-dynamics," that is, the physics of heat.[1] Thomson noted how Hermann von Helmholtz, "adopting the nebular hypothesis," showed in 1854 that the mutual gravitation within the original nebula would have generated heat—enough to explain "the present high temperature of the Sun." The Sun should continue to be hot for "for several million years." Nonetheless, Thomson said, a slow cooling is taking place everywhere, and the current heat of the Sun is in large part a reflection of its "thermal capacity" being "enormous." It follows that smaller bodies like Earth cool faster.

What, then, Thomson asked, does this mean for life? Because at one point, the Earth must have been barren: "Tracing the physical history of the

---

[1] Quotations from Thomson are from "Address of Sir William Thomson, Knt., LL.D., F.R.S., President," especially p. xcix onward.

A Universe of Earths. Dennis Danielson and Christopher M. Graney, Oxford University Press.
© Oxford University Press (2025). DOI: 10.1093/9780197803547.003.0011

Earth backwards, on strict dynamical principles, we are brought to a red-hot melted globe on which no life could exist." Thomson broadly agrees here with Kant's idea that Earth has existed for some time "before it was in a condition to be able to support human beings, animals, and plants." Thomson continues:

> Hence when the Earth was first fit for life, there was no living thing on it. There were rocks . . . , water, air all round, warmed and illuminated by a brilliant Sun, ready to become a garden. Did grass and trees and flowers spring into existence, in all the fulness of ripe beauty, by a fiat of Creative Power?

That explanation won't do, Thomson said. While he would end his speech by affirming "one ever-acting Creator and Ruler," Thomson said that invoking that Creator's power to explain life was not the role of science. "Science is bound," he said, "by the everlasting law of honour, to face fearlessly every problem which can fairly be presented to it. If a probable solution, consistent with the ordinary course of nature, can be found, we must not invoke an abnormal act of Creative Power."

Nor would science allow for the traditional scientific explanation for the existence of life—the "hypothesis of spontaneous generation," as Thomson called it. This was the idea that, under the right conditions, life emerges naturally from inanimate matter.

Genesis may describe life being created by God, who instructs living things to "be fertile and multiply" and fill the seas and land (Gen. 1:22, 28), but for much of history, scientists thought life came from the Earth itself. This idea was ancient, going at least as far back as Aristotle 2500 years ago. This theory of "spontaneous generation" stated that matter possesses an inherent life-generating force. Small creatures could be spontaneously generated from dirt, especially when warmed by the Sun.

Indeed, people claimed to observe this happening. There were reports of frogs seen forming from mud. Ancient Roman writers described this idea. Aelianus (as reported by Edward Topsell) said

> that as he travelled in Italy near Naples, he saw many different frogs by the road, whose fore-parts and head did move and creep, but whose hind-parts were formed and like to the slime of the Earth. This caused the poet Ovid to write thus:

> *Dirt has his seed engendering frogs full green,*
> *Yet so as feetless and without legs on earth they lie,*
> *So as a wonder unto passengers is seen,*
> *One part has life, the other earth full dead is nigh.*[2]

Pliny (also noted by Topsell) also spoke of the regeneration of frogs:

And a strange thing is seen in frogs. After they have lived some six months, they dissolve into a slime or mud—no one knows how. And afterward, with the first rains in the spring, they return again to their former state, in just the same way as they were once first shaped. What their first shape came from, no one knows. The way this happens is also unknown, secret, and beyond comprehension. Yet ordinarily it happens every year.[3]

The Mishnah, the ancient Jewish scripture commentary, discusses whether dirt that generates a mouse would be unclean, since Leviticus 11:29 declares mice unclean: "In the case of a mouse that grows from the ground and is half-flesh half-earth, one who touches the half that is flesh is impure; one who touches the half that is earth is pure."[4]

The ancient Roman poet and philosopher Lucretius said that all sorts of creatures grew straight from the Earth. (He also claimed that the Universe itself formed spontaneously from an infinitude of randomly moving elementary particles, "atoms," and that "other-where the busy atoms join, as well as here; *Such* Earths, *such* Seas, *such* Men, *such* Beasts arise, all like to those enclosed by *our* Skies."[5]) His contemporary, Diodorus of Sicily, described this process, writing of how, with the sea and land arising from the settling elements of the forming (two-storey) Universe,

as the sun's fire shone upon the land, it first of all became firm, and then, since its surface was in a ferment because of the warmth, portions of the wet swelled up in masses in many places, and in these pustules covered with delicate membranes made their appearance. Such a phenomenon can be

---

[2] Topsell, *History of Four-footed Beasts and Serpents*, 719–20. See also Leonhard, "Pictura's Fertile Field," 102–103.

[3] Pliny, *Historie of the World*, Bk. 9, Ch. LI, 265. See also Leonhard, "Pictura's Fertile Field," 103–04 and Topsell, *History of Four-footed Beasts and Serpents*, 720.

[4] Mishnah Chullin 9.6. Translation from https://www.sefaria.org/Mishnah_Chullin.9.6?lang=bi&with=all&lang2=en. See also Leiman, "R. Israel Lipshutz and the Mouse That Is Half Flesh and Half Earth."

[5] Lucretius, *De Rerum Natura*, Bk. II, 65.

seen even yet in swamps and marshy places.... And while the wet was being impregnated with life by reason of the warmth in the manner described, by night the living things forthwith received their nourishment from the mist that fell from the enveloping air, and by day were made solid by the intense heat; and finally, when the embryos had attained their full development and the membranes had been thoroughly heated and broken open, there was produced every form of animal life.[6]

That *every form of animal life* included monsters—unsuccessful life forms. Even human beings could be formed in this way. Today, however, the Earth only generates small creatures like frogs and mice: "the earth constantly grew more solid through the action of the sun's fire and of the winds, [so] it was finally no longer able to generate any of the larger animals, but each kind of living creatures was now begotten by breeding with one another." Or as Lucretius colorfully put it:

> And therefore *Parent-Earth* does justly bear
> The name of *Mother*, since *all* rose from Her:
> She now bears *Animals*, when softning Dew
> Descends; when *Sun* sends Heats she bears a thousand new.
> Well, who can wonder now, that *then she* bore
> Far stronger, bulky *Animals*, and more,
> When *both* were *young*, when both in Nature's Pride;
> *A lusty Bridegroom He*, and *She* a buxom *Bride*?[7]

The "He" in this poem is the Sun, and the "She" is the Earth. The poetic imagery is fantastic. Now imagine this fecund picture of matter applied, not to the traditional two-storey Universe with its single Sun and single world made of earthy matter, but to the sort of Universe envisioned by Copernicus, or better yet, by John Hill. If there are, as Hill put it, many "earthy planets rolling round" other suns, all spontaneously generating every possible sort of living thing, then truly the Universe must teem with life. Indeed, Fontenelle said that, were "the moon but one continued rock" and nothing more, he would even presume it to be populated with rock-eating creatures, for life is everywhere.[8]

---

[6] Diodorus, "The Library of History," Bk. 1:7, 27.
[7] Lucretius, *De Rerum Natura*, Bk. V, 164.
[8] Fontenelle, *Conversations*, 77.

Unfortunately for this picture, spontaneous generation did not withstand scientific scrutiny. Antony van Leeuwenhoek, the pioneering microscopist of the seventeenth century, complained of "respectable and learned men" who told him that eels were spontaneously generated from dew, "in confirmation of which they add, that if no dew has fallen, there will be no eels found."[9] But when observing with his microscopes tiny eel-like creatures that could survive in water mixed with vinegar, he found himself persuaded that even these low, low animals were not spontaneously generated. He reported on this observation in an October 1676 letter to the Royal Society:

> I observed one big drop of water almost from day to day. And after the lapse of about 2 or 3 weeks, I saw that the little eels in this mixed water were greatly increased. Where at first I had seen but 10 eels, I now saw fully 200 of them. And among the rest I saw a great number of very little eels. . . . Seeing this multitude of little eels, I imagined that surely they were not generated from any particles which might have been in the water, nor from any which might have been in the vinegar . . . but I felt firmly persuaded that these little eels had increased in number by procreation.[10]

Studies by van Leeuwenhoek and others undermined the idea of spontaneous generation. By the end of the nineteenth century, the work of scientists like Louis Pasteur would kill off the idea entirely. It did not die easily, however. As late as the 1880s, the German botanist Carl Nägeli was arguing that simple living creatures were spontaneously generated all the time and that higher life forms then evolved from those simple life forms. Thus, the idea of spontaneous generation was still in play when Thomson gave his 1871 address to the BAAS.[11]

Nevertheless, Thomson rejected that "very ancient speculation, still clung to by many naturalists." Mentioning Pasteur's work among others, he insisted that "science brings a vast mass of inductive evidence against this hypothesis of spontaneous generation. . . . Careful enough scrutiny has, in every case up to the present day, discovered life as antecedent to life. Dead matter cannot become living."

---

[9] Leeuwenhoek, "On the Generation of Eels," 62. On eels not procreating being an old idea, see Ruestow, "Leeuwenhoek and the Campaign against Spontaneous Generation," 230.

[10] Dobbell, *Antony van Leeuwenhoek and His "Little Animals,"* 151–53.

[11] Deichmann, "Origin of Life," 343–47; Farley, "The Spontaneous Generation Controversy (1859–1880)," 309; Mason, *A History of the Sciences,* 428–31.

The demise of the stable two-storey Universe and Earth's Copernican elevation to star status had led us to an evolving Universe in which Earth was born as red-hot, dead matter. Presumably all worlds were born this way. So, if life does not naturally emerge from inanimate matter, and if science should not be invoking the Power of God to explain its origin, then what explanation did Thomson propose for its arising on Earth? Or on any other world?

Meteors.

Things will collide in the Universe, Thomson said, and when something large enough collides forcefully enough into a life-bearing planet, bits of material with life on them will be hurled into space. These may eventually fall as meteors on some other planet, seeding it with life in the way life on Earth seeds a newly formed volcanic island.

> Should the time when this Earth comes into collision with another body . . . be when it is still clothed as at present with vegetation, many great and small fragments carrying seed and living plants and animals would undoubtedly be scattered through space. Hence and because *we all confidently believe that there are at present, and have been from time immemorial, many worlds of life besides our own*, we must regard it as probable in the highest degree that there are countless seed-bearing meteoric stones moving about through space. If at the present instant no life existed upon this Earth, one such stone falling upon it might, by what we blindly call natural causes, lead to its becoming covered with vegetation.[12]

Thomson acknowledged that there were scientific objections to this idea and that it "may seem wild and visionary." But the objections are "all answerable"; the idea is "not unscientific" (the contrast here with spontaneous generation or the invocation of Divine Power he left unsaid). He continued:

> From the Earth stocked with such vegetation as it could receive meteorically, to the Earth teeming with all the endless variety of plants and animals which now inhabit it, the step is prodigious; yet, according to the doctrine of continuity . . . all creatures now living on earth have proceeded by orderly evolution from some such origin. Darwin concludes his great work on *The Origin of Species* with the following words:

---

[12] Italics added.

It is interesting to contemplate an entangled bank clothed with many plants of many kinds, with birds singing on the bushes, with various insects flitting about, and with worms crawling through the damp earth, and to reflect that these elaborately constructed forms, so different from each other, and dependent on each other in so complex a manner, have all been produced by laws acting around us. . . . There is grandeur in this view of life with its several powers, having been originally breathed by the Creator into a few forms or into one; and that, whilst this planet has gone cycling on according to the fixed law of gravity, from so simple a beginning endless forms, most beautiful and most wonderful; have been and are being evolved.

With the feeling expressed in these two sentences I most cordially sympathise.

Thus, to sum up Thomson's idea: Earth was once a lifeless red-hot globe, but when it had itself evolved to the point where it was ready to become a garden, it was seeded with some organism that had survived the rigors of being hurled from its native world and transported through space. From that space-seed, life on Earth evolved, into all its elaborately constructed forms, so different from each other.

The idea of space-seeds may have been wild and visionary, but it appealed to others, including Helmholtz, who had also started to advocate for it in the early 1870s.[13] In Helmholtz's view, life was an integral part of the Universe, passing endlessly from world to world. The idea may seem improbable, he wrote in a preface to a translation he made of one of Thomson's books,

But it seems to me a perfectly correct scientific procedure, when all our efforts to produce organisms from lifeless matter fail, to ask whether life ever arose at all, whether it is not just as old as matter, and whether its germs, carried from one celestial body to another, have not developed wherever they have found favorable soil.[14]

The idea of space-seeds thus replaced the idea of spontaneous generation, but the end result, that life was inherent throughout the universe and always had been, was the same.

Svante Arrhenius, the 1903 Nobel laureate for chemistry, wrote in his 1908 book *Worlds in the Making* that

---

[13] On Helmholtz and Arrhenius, see Kamminga, "Life from Space," especially 72–74 and 78–81.
[14] Helmholtz, "Vorrede," xi.

Man used to speculate on the origin of matter, but gave that up when experience taught him that matter is indestructible and can only be transformed. For similar reasons we never inquire into the origin of the energy of motion. And we may become accustomed to the idea that life is eternal, and hence that it is useless to inquire into its origin.

Arrhenius gave the space-seeds idea a catchy name, "panspermia." He also proposed a gentler means of seeding space than collisions: Tiny organisms that had been swept by winds into the upper atmosphere of their planet could be blown out of that atmosphere by the pressure of the light from their planet's star. The light would accelerate them rapidly enough to transport them across the Solar System in months and send them to nearby stars in less than 10,000 years. Experiments with microbes at low temperatures suggested that the cold of space would preserve them.[15]

Arrhenius would go on to argue in 1927 that the Earth was being seeded even now. He thought that thermophilic bacteria (those that live in very high temperatures) could only have evolved on a higher-temperature planet and probably could not sustain a population here over the long term. Therefore, Earth must be being supplied with them, and the most likely source was Venus. Arrhenius cited recent astronomical studies suggesting that the temperature there might be around 50°C (122°F)—perfect for the bacteria he was noting. Organisms blown out of the Venusian atmosphere when Venus was passing between the Earth and the Sun could make it to Earth in mere days, he said.[16]

Thus, following Thomson, Helmholtz, and Arrhenius—and also their twentieth-century successors such as the cosmologist Fred Hoyle (famous for giving the "Big Bang" theory its name as a term of derision and for advocating for an alternative "steady-state" cosmology in which the Universe is eternal with no beginning)[17]—we can conclude that the panspermic Universe is filled with life. Space is filled with the drifting small organisms that are being blown continually into space from planets. Some of these seeds are finding their way to other planets. If the planet that a seed reaches is not suitable to it, it does not thrive; if it is suitable, it does thrive. If the planet is suitable and lifeless, the seed can thrive over time—and through evolution, it can lead to a life-filled world, an "entangled bank," of diverse creatures that

---

[15] Arrhenius, *Worlds in the Making*, 218 (quotation) through 230.
[16] Arrhenius, "Die thermophilen Bakterien und der Strahlungsdruck der Sonne."
[17] Wickramasinghe, "Panspermia According to Hoyle."

can in turn seed other worlds. The result is a Universe in which every planet that possibly can sustain life will host life. The picture is, in some ways, not all that different from the one imagined under spontaneous generation: "many worlds of life besides our own."

Of course, one problem with this picture of a Universe filled with life is that the 1920s' astronomical studies of Venus were wrong. The temperature there is far higher than Arrhenius supposed. Venus is a furnace-like, acid-swept hell-scape. Even Earth's thermophilic bacteria could not survive there. But a greater problem is that of our evolving Universe itself.

12

Cold, Dark Night in a Universe of Galaxies

Life is as old as matter—or so thought Hermann von Helmholtz and Svante Arrhenius. Both men seemed to view matter, and hence life, as eternal. It makes sense, then, according to Helmholz, "to ask whether life ever arose at all." If life is eternal, Arrhenius agreed, then "it is useless to inquire into its origin." These nineteenth-century scientists, who envisioned life being passed from one world in the Plurality of Worlds to another for all time, lived hundreds of years after the demise of the two-storey Universe. Neither would have subscribed to the physics of Aristotle's elements. But they, and much of the scientific world, still hung on to one idea inherited from that two-storey picture: that the Universe is eternal.

Yet the eternity of the Universe was already a dying idea when Helmholz and Arrhenius were pursuing their science. We have seen how already in the eighteenth century Kant recognized that a Copernican Universe governed by Newton's physics must in some sense be *evolving*. Further recognition of that evolving Universe emerged from the study of things we encounter daily—the dark of night, and heat and cold.

The question of the darkness of night arises when we exchange an Aristotelian two-storey Universe—its stars made from an unchanging, eternal, ethereal element and fixed into a big ethereal sphere—for a Copernican Universe with its stars made from normal matter, sitting in empty space, and governed by Newtonian gravity and Newtonian principles of motion. Newtonian gravity says that each star will feel a force of gravitational attraction toward every other star. Two stars may be an enormous distance apart— the gravitational attraction may be minuscule—but Newtonian principles of motion say that if a star in space feels a force, even a tiny force, it will not remain at rest; it will move in the direction of that force. Thus, the consequence of a Newtonian Universe is not just that some sort of collapse to form stars takes place, as Kant noted, but also that stars collapse in upon each other.

Such change presents a problem for an eternal Universe without a beginning. No matter how slow the collapse is, no matter whether the stars move

A Universe of Earths. Dennis Danielson and Christopher M. Graney, Oxford University Press.
© Oxford University Press (2025). DOI: 10.1093/9780197803547.003.0012

toward each other at a rate of a millimeter every million years, if the Universe is eternal, infinitely old, then plenty of time has passed for the collapse to happen. There should be no stars in the night sky. They should all be piled together in a gravitationally bound ball, with the Sun (and its devastated planets) bound right in there with them. And we should not be here.

If the Universe has a definite age, however, this is not a problem. The stars are so far apart that the attraction between them is unimaginably weak, and any collapse has just barely gotten started. Isaac Newton spoke to this in the General Scholium of his *Principia*:

> This most beautiful system of the sun, planets, and comets, could only proceed from the counsel and dominion of an intelligent and powerful Being. And if the fixed stars are the centres of other like systems, these, being formed by the like wise counsel, must be all subject to the dominion of One . . . and lest the systems of the fixed stars should, by their gravity, fall on each other mutually, he hath placed those systems at immense distances one from another.[1]

The collapse might happen—will happen—but not for a very long time.

One way to dodge the collapse problem entirely is simply to say that the Universe is not only *infinitely old*, but also *infinitely large*, containing infinitely many stars. There is then no center toward which any collapse can occur. Every star is surrounded by other stars in all directions and thus is pulled in all directions. The pulls all cancel each other out, and nothing moves. Collapse problem solved! But if the Universe is infinitely large, with infinitely many stars, then you should be able to see infinitely many stars in the sky—and the night sky should not be dark.

Once again, if the Universe has a definite age, then the dark sky issue is not a problem, even if the Universe is infinitely large. In this case, even if there are an infinite number of stars, then light from most of those stars will not yet have reached Earth because the Universe has only existed for a limited amount of time, and light travels at a limited speed. Interestingly enough, this was pointed out in 1848 by the American poet Edgar Allan Poe, who in his prose poem *Eureka* criticized astronomical speculations about size:

> No astronomical fallacy is more untenable, and none has been more pertinaciously adhered to, than that of the absolute illimitation of the

---

[1] Newton, *Mathematical Principles*, 388–89 ("General Scholium").

Universe of Stars . . . observation assures us that there is, in numerous directions around us, certainly, if not in all, a positive limit. . . . Were the succession of stars endless, then the background of the sky would present us an uniform luminosity, like that displayed by the Galaxy—since there could be absolutely no point, in all that background, at which would not exist a star. The only mode, therefore, in which, under such a state of affairs, we could comprehend the voids which our telescopes find in innumerable directions, would be by supposing the distance of the invisible background so immense that no ray from it has yet been able to reach us at all. That this may be so, who shall venture to deny? I maintain, simply, that we have not even the shadow of a reason for believing that it is so.

The paradox of a dark night sky and astronomers' assumptions about an infinitely large, eternal Universe came to be known as Olbers' Paradox, after H. W. M. Olbers, who set it forth in 1826.[2]

Just as common as our experience of the darkness of night is our experience of heat and cold. How we measure these might suggest something unexpected. Consider most of the measurements you make. Can a length be less than zero? No. And while we measure length in different units, like inches, centimeters, and light-years, nevertheless zero inches equals zero centimeters equals zero light-years. This is likewise for time: we cannot have fewer than zero seconds of time, and zero seconds equals zero hours equals zero years. Zero gallons equals zero pints equals zero milliliters.

Now think of measurements of hot and cold. It seems that temperature *can* be less than zero. Moreover, the commonly used temperature scales, Celsius/Centigrade and Fahrenheit, don't even agree with each other on their zero point. Zero Celsius is *not* zero Fahrenheit. Their zeroes therefore cannot be true zeroes.

There actually is a true zero temperature. The study of heat (something prompted by the development of the steam engine and the desire to make such engines more efficient) led to the recognition that heat, once viewed as a kind of invisible "caloric fluid" that made things hot or cold, is in fact a form of energy. This is part of what is called the *First Law of Thermodynamics.* Temperature is a kind of measure of internal energy in an object. There can accordingly be an *absolute* zero temperature where the object lacks any internal energy at all.

---

[2] Poe, *Eureka*, 100. For an English translation of the core of Olbers' argument, see *The Book of the Cosmos: Imagining the Universe from Heraclitus to Hawking*, ed. Danielson, 294–97.

When two objects at different temperatures are brought together, like a warm lump of clay and a cool lump, heat energy naturally flows from the hotter one to the colder one. The warm lump cools and the cool lump warms until the two reach the same temperature—*thermal equilibrium*. But heat will not naturally flow between two lumps of clay at the same temperature so as to cause one lump to warm and the other to cool. Heat naturally flows only from hot to cold; natural processes move toward thermal equilibrium, never away from it. You'll never get warmed up by huddling over a stove that's colder than you are. This is part of what is called the *Second Law of Thermodynamics*.

The Second Law indicates that the Universe is doomed to thermal homogeneity, as hot things must cool off, cool things must warm up, and eventually everything must reach the same temperature. No glowing stars warming the cool seas of their planets, providing energy for life; just a universal darkness that lasts forever as its last bits of temperature difference approach equilibrium, asymptotically.

Newton himself had some inkling of this predicament. He wrote in his *Optics* that, without "active Principles," all things "would grow cold and freeze, and become inactive" and all "vegetation and life would cease." In his notes he wrote that "the globe of this earth & sea was not eternal."[3] It seems, however, that William Thomson (Lord Kelvin), whom we met in the last chapter, was the first to recognize fully this trend toward thermal homogeneity. In an 1852 paper titled "On a Universal Tendency in Nature to the Dissipation of Mechanical Energy," Thomson wrote that at one time "the earth must have been . . . unfit for the habitation of man" (a point to which he returned to in his 1871 BAAS address) and that in the future "the earth must again be" unfit. Later in 1852, Thomson's colleague, the Glasgow engineering professor William John Macquorn Rankine, provided a synopsis of what Thomson was saying:

> The experimental evidence is every day accumulating, of a law which has long been conjectured to exist,—that all the different kinds of physical energy in the universe are mutually convertible. . . . Thomson has pointed out the fact, that there exists (at least in the present state of the known world) a predominating tendency to the conversion of all the other forms of physical energy into heat, and to the uniform diffusion of all heat throughout all matter . . . until all matter shall have acquired the same temperature.

[3] Snobelen, "Apocalyptic Themes in Isaac Newton's Astronomical Physics," 98, 101.

Consequently, Professor Thomson concludes, so far as we understand the present condition of the Universe, there is a tendency toward a condition in which all physical energy will be in the state of heat, and that heat will be so diffused that all matter will be at the same temperature; so that there will be an end of all physical phænomena.[4]

A name was given to this end: *heat death.* In an 1854 paper, Thomson wrote that "the end of this world as a habitation for man, or for any living creature or plant at present existing in it, is *mechanically inevitable.*"[5] But thermodynamics did not just point to an end; it also indicated a beginning. In that same paper, Thomson also wrote that "purely mechanical reasoning shows a time when the earth must have been tenantless; and teaches us that our own bodies, as well as all living plants and animals, and all fossil organic remains, are organized forms of matter to which science can point to no antecedent except the Will of a Creator." As we have seen, Thomson would later argue that science can't appeal to the Creator to solve its problems, and so it seems that questions of heat led Thomson to panspermia.

It was uncomfortable for scientists to hear the evidence offered by heat and by darkness telling them that the Universe was not eternal. The normal scientific view had been that the Universe is stable and unchanging. Aristotle imagined the Universe to be cyclic and eternal, with no beginning and no end. So did many, many people. Aristotle's Universe was imaginable because in the two-storey Universe the heavens that circled above were indestructible and driven by the love of a single unchanging god who Aristotle reasoned must exist. The Sun rose; the Sun set; it drove the water cycle and the whole cycle of life—eternally.

This idea of an unchanging Universe had obviously been challenged by Copernicus and by the Newtonian physics that followed. Nevertheless, well into the twentieth century, scientists were still expecting the Universe to be eternal. This attitude is reflected in the words written in the early 1930s by the astronomer Arthur Eddington (famous for his work in using the solar eclipse of 1919 to confirm that gravity could bend light, as predicted by Einstein's theory of relativity):

[Going back in time] we come to a time when the matter and energy of the universe had the maximum possible organization. To go back further is

[4] Rankine, "On the Reconcentration of the Mechanical Energy of the Universe," 359.
[5] Thomson, "On Mechanical Antecedents of Motion, Heat, and Light," 61.

impossible. We have come to an abrupt end of space-time—only we generally call it the "beginning". I have no "philosophical axe to grind" in this discussion. Philosophically, the notion of a beginning of the present order is repugnant to me. I am simply stating the dilemma to which our present fundamental conception of physical law leads us. I see no way around it.[6]

The dark of night, and heat and cold, are familiar things that point to a beginning, but arguably the big pointers were nebulae—those things that Kant supposed to be other galaxies. Most nebulae, seen through a telescope, look like dim, fuzzy blobs. In Ireland during the 1840s, William Parsons built what at the time was the largest telescope in the world, in hopes of learning more about nebulae. This telescope, which gathered light with a mirror whose diameter measured 6 feet and which had a tube several times that in length, would be a good-sized instrument even in modern times (the Vatican Observatory's primary telescope today is of similar size). In the 1840s, it was gargantuan. With this giant telescope Parsons observed that many of the nebulae had a spiral structure.

To some, these spirals seemed most likely to be not Kant's other galaxies, but rather solar systems in the process of forming. In 1888 the English astronomer Isaac Roberts obtained a high-quality photograph of the Andromeda nebula (see Figure 12.1) that revealed it to be a spiral. Roberts wrote, "Here we (apparently) see a new solar system in the process of condensation from a nebula—the central sun is now seen in the midst of nebulous matter which in time will be either absorbed or further separated into rings."[7] This was, of course, in line with Kant's thinking. Roberts's work convinced many. Agnes Mary Clerke (1842–1907), who served as a sort of one-woman, multilingual hub for astronomical information in the later nineteenth century, wrote in her influential *The System of the Stars* that

The question of whether nebulae are external galaxies hardly any longer needs discussion. It has been answered by the progress of discovery. No competent thinker, with the whole of the available evidence before him, can now, it is safe to say, maintain any single nebula to be a star system of coordinate rank with the Milky Way.[8]

6 Eddington "The End of the World," 450.
7 Roberts, "Photographs of the Nebulæ M31, etc.," 65.
8 Clerke, *The System of the Stars*, 2nd ed., 259, 349. Regarding Clerke's work and her role as a "hub," see Brück, *Agnes Mary Clerke and the Rise of Astrophysics.*

Figure 12.1 Photograph of the Andromeda "nebula" by Isaac Roberts, 1888. Roberts believed this photograph showed the nebula to be a new solar system in formation. Today we know it to be a relatively nearby galaxy.
Credit: Wikimedia Commons

Apparently, it was *not* safe to say, however, because a little more than thirty years later astronomers were still pondering what sorts of objects spiral nebulae were. In 1920, Harlow Shapley and Heber Curtis had a debate on this question that became famous among astronomers. Neither astronomer would be considered wholly correct in light of today's science. However, Shapley had devised an interesting procedure that bears on our story. He had used a specific kind of star called a "Cepheid variable" to estimate the size of our Milky Way.

Cepheid variables were so named because they cycled in brightness according to a distinctive pattern linked to their overall light output. Their archetype was a star in the constellation Cepheus. An object with known light output, or *luminosity*, can be used to measure distances; once the luminosity is known, distance can be determined based on how bright or dim the object appears from Earth. This procedure works best if the object is highly luminous and thus is easily seen for a great distance. Cepheids are powerful stars.

The nature of Cepheids had recently been discovered by Henrietta Swan Leavitt, who worked as a "computer" for Harvard observatory (much like the "computers" portrayed in Margot Lee Shetterly's 2016 book *Hidden Figures: The American Dream and the Untold Story of the Black Women Mathematicians Who Helped Win the Space Race* and in the movie of the same name). Leavitt had been tasked with cataloguing large numbers of stars from photographs, and in the process of doing so she discovered the nature of Cepheids.

Leavitt's was a discovery of great importance because it allowed astronomers to measure vast distances in space. Any astronomer who could see a Cepheid in some object could time its cycle and deduce its luminosity, and hence its distance. In the 1920s, Edwin Hubble (1889–1953) was able to use a new, large (8.3 feet in diameter) telescope on Mount Wilson in California to find Cepheids in the Andromeda nebula. Leavitt's stars revealed that the Andromeda nebula was far too distant to be in our Milky Way. That nebula is large as seen from Earth—larger even than the full Moon. To be so distant and yet appear so large required that it be of vast size. It was both outside of and "of coordinate rank with" our Milky Way. The Andromeda "spiral nebula" was the Andromeda *galaxy*. The spiral nebulae were other galaxies.

Hubble also showed that the galaxies were receding from each other. The motions of galaxies cannot be observed directly; rather, their motion is inferred from a slight reddening of the light they emit and is best observed in astronomical objects when the light is dispersed by a prism into a rainbow or "spectrum" of color. This effect is observed in objects on Earth and underlies, for example, the radar units used to measure the speed of baseballs and speeding cars. The galactic "red shifts" indicated that the Universe is expanding.

Belgian physicist Georges Lemaître (1894–1966) was also a key figure in the story of the expanding Universe. In the 1920s he spent a year in Cambridge working with Arthur Eddington, and he worked a further year at Harvard with Harlow Shapley. Later in the 1920s, he likewise proposed, based on his reading of Albert Einstein's theory of relativity, that the Universe was expanding.

An *expanding* Universe was of course smaller in the past than it is today, and its expansion also strongly suggests that it is not infinitely old, not eternal, but that it had a beginning. As we have seen, Lemaître's colleague Eddington found such a Universe "repugnant." Hubble likewise became uncomfortable with such a Universe and eventually reversed his position regarding cosmic expansion. By 1942, he was arguing that the reddening he had measured was due to some unknown phenomenon, not to galactic motions:

> Red shifts are due either to recession of the nebulae or to some hitherto unrecognized principle operating in internebular space. . . . [The] interpretation of red shifts as velocity shifts leads to a particular type of an expanding universe which is disconcertingly young, small and dense.

Thus Hubble argued (or conjectured) that what astronomers were observing was "a new principle of nature" in action, not an expanding, young (compared to infinitely old) Universe.[9]

Lemaître, however, stood his ground against colleagues such as Eddington, waxing poetic, even philosophical, in defending his view of the Universe. From one angle this view now appears as the "ashes and smoke of bright but very rapid fireworks," and more positively as "a world where something really happens [since] the whole story of the world need not have been written down in the first quantum like a song on the disk of a phonograph."[10]

Nor was Lemaître's position shaken by other challenges to the idea of an expanding Universe, including some from the Soviet Union. In 1949 Soviet astronomers—dialectical materialists who were mightily bothered by the

---

[9] Hubble, "The Problem of the Expanding Universe," 214–15.
[10] Lemaître, *BOTC*, 408.

fact that Lemaître's other vocation was that of a Catholic priest—pledged to fight against the theory of the "widening of the Universe." They declared that, to counterbalance this allegedly pro-religious "bourgeois" idea, "Soviet science must intensify its work on regions beyond our galaxy, to give a materialistic explanation of the red displacement in the spectra of galaxies."[11] Members of the Soviet establishment thus rejected the expanding Universe:

> Today's bourgeois science supplies the church and fideism with new arguments . . . [which] assert that the world is finite, that it is limited in time and space. . . . The theory ["broadly known in capitalist countries as the expanding Universe theory"] was offered in the late 1920s by the Belgian priest Georges Lemaitre, and it is underlain by the phenomenon called in astronomy the "red shift." The reactionary scientists Lemaitre . . . and others made use of the "red shift" in order to strengthen religious views on the structure of the Universe. . . . Falsifiers of science want to revive the fairy tale of the origin of the world from nothing.[12]

And

> The general scientific crisis in America and Western Europe is reflected in astronomy also. As a result, cosmogony, the branch of astronomy concerned with questions of the development of celestial bodies, has been turned into a repository for all kinds of idealistic nonsense and absurd fabrications which in the last analysis aim at restoring the legend of the world's creation.[13]

Despite these protests, Lemaître's ideas have stood the test of time, and they undergird what is now known (thanks to Fred Hoyle's derisive coinage) as "The Big Bang Theory." Thomson had spoken in 1871 of Earth once being "a red-hot melted globe on which no life could exist." According to the Big Bang theory, the Universe itself was once very small, and far, far hotter than any molten globe. In fact, it was so hot that even atoms could not exist. The remnant heat from that early, hot state is still present within the Universe; astronomers study this "Cosmic Microwave Background" to learn about the Big Bang. The Universe is expanding, and cooling, and has a finite age of

---

[11] Salisbury, "Russian Astronomers Hold Theory of Cosmos Origin Surpasses West."
[12] Tropp et al., *Alexander A. Friedmann*, 223–24.
[13] "Greetings to Winners of Stalin Prizes." The quote is from V. Ambartsumyan, who was a Stalin Prize winner and president of the Armenian Academy of Sciences.

about 14 billion years—in agreement with the dark night (and Poe), with thermodynamics, and with observations of galaxies.

This galactic Universe poses interesting questions regarding other earths. On one hand, many galaxies means many, many more stars and thus more planetary systems than would exist if the Universe consisted of our Milky Way alone. Indeed, it multiplies their numbers by as many times as there are galaxies, and the galaxies number in the hundreds of billions. But a galactic Universe offers no escape from the question of the origin of life.

Thomson stated that Earth initially had to be lifeless because it was too hot. The Universe, however, was once far, far hotter. The Universe is not eternal; neither, then, is life. In fact, life cannot be even as old as the Universe. It cannot precede the time when some part of the Universe first cooled enough for life to exist. Life had an origin *within* the Universe—we are here, after all—and yet life does not spontaneously generate from matter. We do not see life originating from inanimate matter in nature, and we have so far been unable to generate life from inanimate matter in a laboratory. The astonishingly vast but apparently infecund Universe to which Copernicus led us is a puzzle.

13

"Variety Enormously Greater than Had Been Supposed"

We do not expect to find life on Venus, despite the fact that Venus might seem to be an Earth-like planet. Venus has nearly the same diameter as Earth, a comparable mass, and, like Earth, an atmosphere. Also, Venus lies within the "habitable zone," that area surrounding the Sun that is neither so close to it that temperatures would be too hot for life nor so distant that temperatures would be too cold (although Venus rides the inner edge of that zone). Svante Arrhenius imagined Venus, which in his day was thought to have temperatures around 50°C (120°F), as the panspermic source of the heat-loving bacteria found in hot springs here on Earth.

But Venus is not Earth-like. Its temperatures are above 850°F (over 450°C). Its atmospheric pressure is nearly a hundred times that of Earth. The clouds of its atmosphere are full of sulfuric acid. It is a very, very inhospitable place. Were you to be magically transported there, as you are, you would die instantly.

The same holds true for almost any place in the Solar System other than Earth. The Moon, which of course is as much in the habitable zone as Earth, lacks any atmosphere to speak of because its gravity is too weak to hold on to one. The Moon lacks a magnetic field to shield it from particles coming from the Sun and from space. Its surface, therefore, is airless, waterless (contrary to Kepler), scorching where illuminated by the Sun, far colder than the coldest spot in Antarctica where shaded from the Sun, and subject to toxic radiation. If you went to the Moon as you are right now, with no special equipment, no space suit, you would die fast. Mars is also in the habitable zone (although on the outer edge), but it has other problems that would ensure that, were you suddenly to find yourself there, you would not last long.

Mercury is scorching and airless. Jupiter, Saturn, Uranus, and Neptune don't even have surfaces onto which you could be transported—they are balls of gas and ice that just sort of get thicker as you go down into them.

A Universe of Earths. Dennis Danielson and Christopher M. Graney, Oxford University Press.
© Oxford University Press (2025). DOI: 10.1093/9780197803547.003.0013

The wind alone in these gas balls would rip you apart (winds on Neptune top 1000 mph). Pluto, Ceres, and the other small bodies of the Solar System would all have the same sorts of problems as the Moon, even were they in the habitable zone.

Things might go better for you were you transported into the cold, black waters under the ice of Saturn's moon Enceladus or Jupiter's moon Europa. You might last on those moons a little while—that is, if you had the right scuba gear.

The work of Copernicus may have suggested that Earth is "like any other planet" (to borrow words from his sixth Axiom), but clearly our Solar System is not full of other worlds like Earth. And William Whewell (1794–1866), a polymathic poet and mathematician who also served as Master of Trinity College, Cambridge, may have been the first person to envision the sort of Solar System we've just described. It was Whewell who in 1834 coined the word *scientist* as a catch-all term to denote those active in the different disciplines of science. And then in 1853, as a scientist, he published—anonymously—*Of the Plurality of Worlds: An Essay*, attacking the idea of a Universe widely inhabited with intelligent life.[1]

Before Whewell it had been common to assume that the Solar System's planets were Earth-like and inhabited. Huygens had argued that a Copernican would presume that "the rest of the planets have their dress and furniture, nay and their inhabitants too as well as this earth of ours." Huygens presumed that, since Jupiter had clouds, it would also have oceans, and thus sailors, and sailing ships—and all the things found on sailing ships, too: "If they [the Jovians] have Ships, they must have Sails and Anchors, Ropes, Pullies, and Rudders . . . and perhaps they may not be without the Use of the Compass too."[2]

The great William Herschel (1732–1822), who in the late eighteenth century became the first astronomer to discover a new planet (Uranus), even proposed that the Sun itself was inhabited, "like the rest of the planets." He surmised from the dark spots on the Sun "that its surface is diversified with mountains and valleys." And so, he continued (arguing from analogy),

this way of considering the sun and its atmosphere, removes the great dissimilarity we have hitherto been used to find between its condition and that of the rest of the great bodies of the solar system.

---

[1] Crowe, "William Whewell." See also Danielson, "Scientist's Birthright."
[2] Huygens, *Celestial Worlds Discover'd*, 83.

The sun, viewed in this light, appears to be nothing else than a very eminent, large, and lucid planet, evidently the first, or, in strictness of speaking, the only primary one of our system; all others being truly secondary to it. Its similarity to the other globes of the solar system with regard to its solidity, its atmosphere, and its diversified surface; the rotation upon its axis, and the fall of heavy bodies, leads us on to suppose that it is most probably also inhabited, like the rest of the planets, by beings whose organs are adapted to the peculiar circumstances of that vast globe.... *Upon astronomical principles*, [I] propose the sun as an inhabitable world, and am persuaded that the foregoing observations, with the conclusions I have drawn from them, are sufficient to answer every objection that may be made against it.[3]

Given that in Herschel's time life was supposed to generate spontaneously from matter, and given the old idea that matter generated all sorts of life forms and that those suitable to the environment survived (with the "monsters" dying off—recall from Chapter 11), Herschel's idea of an inhabited Sun had a certain logic to it. He was not the only one to imagine life on the Sun. François Arago, Director of the Paris Observatory from 1843 to 1853, likewise proposed that the Sun was inhabited. Thomas Dick in the 1830s estimated the population of the Solar System at over twenty trillion, not counting "innumerable orders of sentient and intelligent beings" on the Sun.[4]

In *The Age of Reason* (1794), Thomas Paine had written of the inhabitants of "each of the worlds of which our [solar] system is composed" and of those "millions of worlds" around other stars.[5] Jérôme de Lalande, senior director of the Paris Observatory in the late eighteenth century, felt it was foolish to think the Solar System uninhabited. Yet again based on analogy, he declared:

The resemblance between the earth and the other planets is so striking, that if we allow the earth to have been formed for habitation, we cannot deny that the planets were made for the same purpose.... there is every possible resemblance between the planets and the earth: is it, then, rational to suppose the existence of living and thinking beings is confined to the earth? From what is such a privilege derived but the groveling minds of persons who can never rise above the objects of their immediate sensations?[6]

---

[3] Herschel, "On the Nature and Construction of the Sun and the Fixed Stars," 63.
[4] Crowe, *The Extraterrestrial Life Debate, 1750–1900*, 246–47 (Arago), 198–99 (Dick).
[5] Crowe, *The Extraterrestrial Life Debate: Antiquity to 1915*, 228–29.
[6] Lalande, "Introductory Note," vi–vii.

Whewell was therefore venturing into what some considered the realm of the irrational and the groveling—yet with science! Historian of science Michael J. Crowe points out that a key element of Whewell's argument was his application of the inverse-square law as it relates to gravitational force, light, and heat radiation. This law says that if planet B is twice the distance from the Sun as planet A, it will receive one quarter the heat and light from the Sun, all else being equal. Accordingly, were Earth a little closer to or further from the Sun, then the amount of heat and light it would receive would be significantly different. Crowe argues that Whewell was one of the first to consider the idea of a habitable zone around a star (an idea that of course implies that the rest of the region around a star is *not* habitable)—a concept he put forward in the *Essay* using the term *temperate zone*.[7] Astronomers today also sometimes refer to the "Goldilocks zone," the zone that is not too hot and not too cold.

Whewell also used geology to argue against widespread extraterrestrial intelligence. As Crowe notes:

> Whewell's argument was that evidence for the age of the Earth showed that throughout most of Earth's history it had been bereft of intelligent life, which suggested that the Creator's plan for the cosmos was capacious enough to leave vast regions of it lacking [intelligent beings] for long periods of time.[8]

Given that Whewell was president of England's Geological Society, the argument was an authoritative one.

Whewell did not rule out the idea of lower life forms within the Solar System. For example, he speculated about life existing on Jupiter (which for various reasons he thought would be a water world), writing that Jovian life forms might be "aqueous, gelatinous creatures; too sluggish, almost, to be deemed alive, floating on their ice-cold water, shrouded forever by their humid skies."[9]

Crowe notes that the scientific evidence for diverse planets, and not merely other earths, had been available long before Whewell. The great difference in the sizes of Earth and Jupiter had been known since Huygens (recall Figure 1.2). Newton had included in the third edition of his *Principia*

---

[7] Crowe, "William Whewell," 438–41.
[8] Crowe, "William Whewell," 441.
[9] Whewell, *Of the Plurality of Worlds*, 185–86.

(1726) calculations showing the substantial differences in the masses, densities, and surface gravities of different Solar System bodies. William Whiston, Newton's successor to the Lucasian Chair of Mathematics at Cambridge in 1701, showed a wide range of differences in the light and heat received by planets in the "Solar System *Epitomis'd*" broadside he published in 1712 (see Figure 13.1). Such differences might have provided a reason to suppose that the planets were sufficiently unlike our Earth to be devoid of intelligent life.[10]

Astronomers prior to Whewell did recognize planetary diversity. However, that recognition was limited and was inclined nevertheless toward viewing planets as habitable. Consider Whewell's friend John Herschel, son of William Herschel, and one of the more prominent astronomers of Whewell's time. Prior to Whewell's 1853 *Essay*, John Herschel had discussed how much more solar heat Mercury receives than Uranus, how much stronger gravity must be on Jupiter than on Earth, and how the density of Saturn is so much less than the density of Earth that Saturn "must consist of materials not much heavier than cork."[11] Yet he still presumed that the planets would be inhabited, much as his father presumed the Sun would be.[12] Indeed, he marveled:

> What immense diversity must we not admit in the conditions of that great problem, the maintenance of animal and intellectual existence and happiness, which seems, so far as we can judge by what we see around us in our own planet, and by the way in which every corner of it is crowded with living beings, to form an unceasing and worthy object for the exercise of the Benevolence and Wisdom which presides over all![13]

When Whewell's objections to a Plurality of Worlds were published, Herschel wrote to him that, "though somewhere I have myself stated that taken in a lump Saturn might be regarded as made of Cork—it *never did* occur to me to draw the conclusion that *ergo* the *surface* of Saturn must be of extreme tenuity." Herschel then went on to propose that even were Jupiter an ice-cold sea, it might be populated by something more than Whewell's gelatinous

---

[10] Crowe, "William Whewell," 432–35.

[11] Herschel, *A Treatise on Astronomy*, 263.

[12] For example, Herschel, in *A Treatise on Astronomy*, refers to inhabitants on Mars (264) and Saturn (411).

[13] Herschel, *A Treatise on Astronomy*, 263.

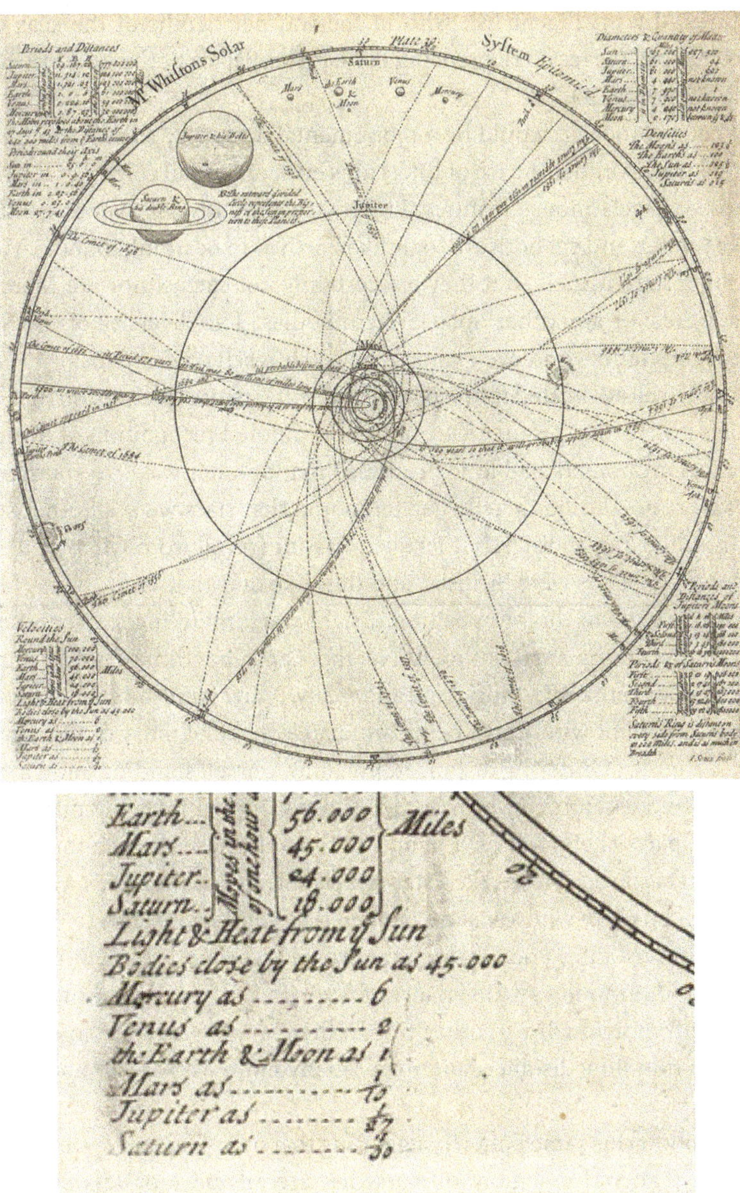

Figure 13.1 William Whiston's 1712 "Solar System *Epitomis'd*" (top) includes, in the lower left corner, a section (bottom) showing how the planets receive significantly differing amounts of light and heat from the Sun.

Credit: Image courtesy of Barry Lawrence Ruderman Antique Maps, Inc.

creatures—by intelligent fish who in the warmer depths of the seas construct "crystal palaces" (referencing the Crystal Palace of the London Great Exhibition of 1851).[14]

Of course, today it would be a monumental discovery were we to find in the cold waters under the ice of Enceladus even one small colony of the most microscopic gelatinous creatures. Planets truly are diverse: Venus is utterly unlike Jupiter, and yet both are so unlike Earth as to be uninhabitable. Venus and Jupiter are planets, but they are certainly not earths; nor are Mercury, Mars, Ceres, or any other Solar System bodies. The diversity of the Solar System goes far beyond even what Whewell himself envisioned.

As Whewell advocated for recognizing the diversity of planets, so the work of other nineteenth-century astronomers promoted recognition of diversity in stars. Astronomers had long presumed that one star was like another. Of course, Bruno and Fontenelle had supposed that stars were all other suns, and thus like one another, but Jacques Cassini (recall from Chapter 9) had supposed that stars were Siriuses, and thus like one another.

There is logic to this supposition. Bright stars, middling stars, and faint stars don't come in equal numbers. No, there are a handful of bright stars in the night sky, more middling stars, and many, many faint stars. This appearance is consistent with Earth being amid a Universe of more or less similar stars, scattered at differing distances. Imagine yourself standing amid a large but scattered assembly of people, like those gathered for a big public fireworks display. There will be a handful of people close enough to you that you might easily be able to take a step, reach out, and shake hands with them. There will be more who are close enough that you might shout to them and they might recognize your voice in the crowd and wave back. But there will be thousands around you too far away for you to really recognize or communicate with individually. In other words, there are few people close to you, more at a middling distance, and many, many more who are far away in the crowd.

Stars follow this same pattern, suggesting that our planetary system stands amidst a scattered assembly of stars that are all more or less similar to one another (just as people are more or less similar—none of us is 50 feet tall). Were stars not at differing distances, but instead on a sphere and thus equidistant from Earth, they could come in any distribution of brightnesses (in which case those brightness differences would be inherent in the stars,

---

[14] Herschel, "Letter to W. Whewell of 3 January 1854," 358–59.

and not a factor of distance). There could be as many bright stars as faint stars.

Back in the time of the two-storey Universe, plenty of astronomers did assume that stars were on such a sphere, but others looked at the stars and assumed the scattered-distance model. Albert the Great, for instance, wrote in the thirteenth century how "We ourselves have indeed recognized one star to be more or less distant than another through the greater or lesser [apparent] diameter of the stars: but we are not able to distinguish the number of stars."[15] Of course, we have seen how scientists like Thomas Digges adapted the scattered-distance model for the Copernican Universe.

Nevertheless, all these astronomers were merely making assumptions. Science provided no positive information concerning the stars and their distances. Toward the end of the nineteenth century, Agnes Mary Clerke, that one-woman information hub we met in Chapter 12, described what was known about the stars at that century's start and what had been learned about them by its end.

At the start of that century, Clerke wrote, very little was known; astronomers largely viewed the stars as a background against which the motions of planets were measured. Even in the last quarter of the eighteenth century, she wrote, "the sidereal world . . . was the domain of far-reaching speculations," unencumbered by systematic study.[16] William Herschel, she noted, operated on the assumption that "the brightness of a star afforded an approximate measure of its distance"—assuming that one star was more or less the same as another, so those stars that appeared faint to us he reckoned to be more distant, and those that appeared bright to us, less distant.[17] Therefore, the stars might be (more or less identical) suns. Nothing stood against this assumption, since so little was known about the stars, but neither was there any particular science to support it.

Through the nineteenth century, however, more and more was learned about the stars, thanks largely to advances in technology. Astronomers finally developed telescopes of sufficient quality to be able to determine stellar distances through measurements of parallax. By the end of the century, Clerke wrote, distances to roughly one hundred stars had been measured.

[15] Magnus, *Commentarii*, Dist. 2, Art. 3, "Ad Quaest.": "nos quidem deprehenderemus unam stellam magis vel minus distare quam aliam per diametrum stellarum magis vel minus: sed non possumus distinguere numerum stellarum."
[16] Clerke, *A Popular History*, 11.
[17] Clerke, *A Popular History*, 21.

Moreover, "the list [of stars with measured distances] is an instructive one, in its omissions no less than in its contents. It includes stars of many degrees of brightness, from Sirius down to a nameless telescopic star in the Great Bear."[18]

Clerke goes on to point out something very interesting and highly significant: many of the brightest stars in the sky had been found to be too far away for their distances to be measured; meanwhile, most of the stars that were found to be nearest to Earth were quite faint. And so, she wrote,

> The obvious conclusions follow that the range of variety in the sidereal system is enormously greater than had been supposed, and that estimates of distance based upon [brightness seen from Earth] must be wholly futile. Thus, the splendid Canopus, Betelgeux, and Rigel can be inferred, from their indefinite remoteness, to exceed our sun thousands of times in size and lustre; while many inconspicuous objects, which prove to be in our relative vicinity, must be notably his inferiors. The limits of real stellar [luminosity] are then set very widely apart.[19]

In other words, stars are definitely not all suns. But we certainly understand today that the Sun is a star: that it is a gravitationally bound globe of dense gas, heated to incandescence by nuclear reactions occurring deep within it. And we understand that all stars are nuclear-powered, gravitationally bound, incandescent gas globes.

As Clerke observed, however, the range of what counts as a "star" is very wide. We have learned more about more stars in the century-plus since she wrote, but her observation still holds true. There actually are stars that are enormous, comparable in size to the sizes Kepler determined (but not at all dim); these are rare. More common are stars comparable to the Sun. But most common of all are those stars "notably inferior" to the Sun. Of the hundred stars currently known to be the Sun's nearest neighbors in space, roughly eighty have less than one one-hundredth of the Sun's power output.[20] All these combined would not equal the Sun. Most of these stars are not even visible to the naked eye.

---

[18] Clerke, *A Popular History*, 37.

[19] Clerke, *A Popular History*, 37.

[20] Based on "Star Browser" data from Celestia 1.6.2.2 (2001–2021) and the Research Consortium On Nearby Stars (Georgia State University) list of "The One Hundred Nearest Star Systems": http://www.astro.gsu.edu/RECONS/TOP100.posted.htm.

Following the downfall of Kepler's and Cassini's star size arguments, science was neutral on the question of whether stars were other suns. But in the nineteenth century science came back to show that Kepler and Cassini were right conceptually, if wrong on the particulars: stars are not merely other suns.

Again, recall humans' love of analogy, evinced repeatedly since our citation of John Hill's *Urania* and discussion of Kant's theory of the nebulae-as-other-galaxies. There we noted both the power and the limitations of analogy, citing L. S. Stebbing's caution that "the greater our ignorance of the subject-matter, the more likely we are to be misled by a weak analogy." In her work, Clerke illustrated how nineteenth-century science relentlessly undermined the analogy of "stars as other suns" that had been presumed in the absence of any real knowledge. The scientific picture was increasingly not of analogy but of *dis*analogy among stars.

More recently, other evidence has emerged of the non-likeness of star with star: For example, among a sample of 369 stars that *are* in many respects similar to the Sun in age, temperature, surface gravity, metallicity, and period of rotation, there appear *five times* the variation in brightness of that displayed by our Sun—a variability that can profoundly affect life-critical features such as the atmospheric chemistry and the climate of a planet orbiting a given "sun."[21] Even among similar stars, therefore, our very-stable Sun is not just another star.

There is great diversity among stars, just as there is great diversity among planets. The two-storey Universe may be gone, and both Earth and Sun may truly be participants in the "dance of the stars." But these and the other dancers may differ from one another far more than we ever could have expected when we first imagined Earth as a star.[22]

---

[21] Reinhold et al., "The Sun Is Less Active than Other Solar-like Stars."

[22] Further discussion of Whewell and Clerke and the matter of diversity among planets and stars can be found in Christopher M. Graney, "The challenging history of other Earths," *International Journal of Astrobiology*, 22:6 (2023): 729–38, and Christopher M. Graney, "The Challenging History of New Worlds: How a revolutionary idea was born and gained acceptance despite a lack of solid scientific support," *Sky & Telescope* (August 2024): 28–33.

14

Junctions on the Road to *Star Wars*

The work of Whewell and of Clerke established the thesis of cosmic diversity: diversity both among planets and among stars. In this way their contributions seem to mark the second of two major historical junctions in the road to the Plurality of Worlds—to the worlds of *Star Wars*, Marvel, and so many other popular, fictional renderings of ETI. Both junctions in that road suggest that belief in other earths is not driven and supported by scientific discoveries but instead arises and is sustained simply because the idea is so marvelous, so alluring, so "universally interesting."

We take this last phrase from one of William Whewell's most prominent critics, the Scottish physicist David Brewster. Brewster was another proponent of life on the Sun; according to him, "Universal life upon universal matter is an idea to which the mind instinctively clings." In his *More Worlds than One* (1854, 1870), Brewster declared that "there is no subject within the whole range of knowledge so universally interesting as that of a Plurality of Worlds"[1]—and who could disagree? It certainly retains that interest and fascination today.

Brewster saw no junctions at all on the road to a Plurality of Worlds. He dismissed Whewell's case, alleging it to be based on "conjecture" and "insulting to Astronomy." Like Jerôme de Lalande, Brewster associated any questioning of Pluralism with mental deficiency, offering the condescending suggestion that Whewell's arguments must be ascribed "only to some morbid condition of the mental powers, which feeds upon paradox and delights in doing violence to sentiments deeply cherished and to opinions universally believed"[2]—echoes of Lalande's "groveling minds" aspersion. Brewster saw the road as an uninterrupted, straight highway: no junctions, no bumps, and those who thought otherwise needed to have their heads examined.

Yet the first junction had appeared already in the seventeenth century, with the work of Kepler. We have seen how, before Copernicus and Kepler

---

[1] Crowe, *The Extraterrestrial Life Debate, Antiquity to 1915*, 356
[2] Crowe, *The Extraterrestrial Life Debate, Antiquity to 1915*, 356–57.

A Universe of Earths. Dennis Danielson and Christopher M. Graney, Oxford University Press.
© Oxford University Press (2025). DOI: 10.1093/9780197803547.003.0014

and Newton, those who studied the Universe thought of it as having two storeys. It was also thought of as stable, eternal, and a beautiful blue. Earth was the lowest thing in it, but life sprang forth even from the mud of the Earth, and sprang forth all the time.

Copernicus, in trying to explain the motions of the wandering stars, proposed that our Earth is a wandering star within the Universe, like Mars and Venus. At first, the Copernican Universe retained much of the old two-storey Universe, simply putting the core of that Universe into orbit around the Sun—as exemplified brilliantly by Thomas Digges in the sixteenth century. However, a Universe in which Earth and Mars have something in common does not sit well with the ideas of "up there" and "down here," and astronomers soon began seeing similarities, *analogies*, between there and here, aided by telescopic observations.

Some of those similarities were real: Mars and Venus do circle the Sun, like Earth; like Earth, too, they are round. And some of that analogical thinking paid off. Mars and Venus are indeed made of the same matter as Earth and are governed by the same laws of gravity and motion. However, some of the apparent similarities did not withstand further scrutiny, and some of the analogical thinking did not pay off. There are no lunar lakes like those cited by Kepler as evidence that the Moon, like Earth, had gravity. Bruno and his successors extended the analogies (without support from empirical evidence) across the Universe, envisioning a Plurality of Worlds, so that the Sun together with its system of planets was but one among many, and those planets were other earths.

The first junction point came when Kepler adamantly pointed out that the available science of his time implied that stars in a Copernican Universe were *not* other suns. Kepler's science was seemingly solid. Any observer could verify what he was saying. The idea of Copernican stars being suns arguably should have withered at that point, and should have remained withered, through at least the middle of the seventeenth century. Kepler's data told him that the Sun was a unique body in the Universe, and thus its planetary system was unique, and so was our Earth. At this point science plausibly could have, and arguably should have, turned off the road to Pluralism. The Keplerian Universe of one tiny but brilliant Sun amid many giant but dim stars (that were not at all suns) could plausibly have become the accepted model for quite a long time.

Eventually, of course, the problems with the measurements of star size that underpinned Kepler's Universe would have been exposed. Huygens would have made his own, more accurate measurements. But the notion of stars as suns would then have had to fight its way through the supporters of the accepted Keplerian model. Recall that Huygens's claims about star sizes did not themselves go unchallenged. When Huygens alleged that the telescope was producing spuriously large images of stars, John Flamsteed, England's first Astronomer Royal, invoked telescopic observations of Mercury and Sirius to argue against him. The telescope shows the globe of Sirius even better than that of Mercury, Flamsteed stated. Had Kepler's "single Sun" cosmology already been broadly accepted, it would not have been displaced easily.[3]

But no turn was made at Kepler's junction. By the second half of the seventeenth century, people found Fontenelle's ideas of a Pluralistic Universe compelling, despite contrary and easily reproducible telescopic data. And by the start of the nineteenth century, astronomers and others regarded the Universe as a place full of other suns, orbited by other earths. Indeed, every planet was an earth, those like Mars and Venus that orbited our Sun, as well as all those other planets orbiting other suns. Some, as we've noticed, even saw the Sun itself as a home to sentient life. Our Solar System teemed with life. Yet the Pluralistic Universe of that time, so different and so much larger than the old two-storey Universe, retained (acknowledged or not) certain important aspects of the old two-storey model of the Universe: It was stable, essentially eternal; and life naturally sprang forth spontaneously from its matter.

At the end of the nineteenth century, however, astronomy reached a second junction, and everything upon which the Pluralistic Universe was based was in jeopardy. Science had shown that life did not spontaneously spring forth from inanimate matter, and indeed that the Universe was not stable; instead, it was evolving, and in its earlier stages Earth itself was an environment hostile to life. Science had revealed that planets were not simply other earths; they were diverse bodies. Stars were not simply other suns; they too were diverse. The Universe that William Shatner saw in 2021, dark and lifeless beyond Earth, the Universe of our pale blue dot in the inky blackness, was already on the horizon at the end of the nineteenth century. And since then, that Universe has decisively arrived from over the horizon. Today's science simply accentuates what was then coming into view.

[3] Graney, *Setting Aside*, 148-58.

As we have seen, today's science reveals the evolutionary nature of the Universe in a way that raises difficult questions regarding life itself. Since before Aristotle we had presumed that the existence of life went hand in hand with matter. Brewster (again) agreed: "wherever there is matter there must be life"—a perfect fit to Herschel's Sun-dwellers and Fontenelle's lunar rock-eaters. But life does *not* spontaneously emerge from inanimate matter: Science has never observed life arising from such matter, neither in nature nor in the laboratory. Moreover, under the Big Bang Theory, the Universe began in a state so hot that even atoms could not exist, let alone life. Life therefore had to arise from inanimate matter, but (obviously) through a process that does not occur regularly and that continues to elude our understanding. Never in the history of science have we been so in the dark as we are today—or, perhaps, so cognizant of how in the dark we are—regarding how life came into existence.

Today's science reveals diversity in the Universe that also raises difficult questions regarding what sort of planet could truly *be* another Earth. The discovery of exoplanets—planets orbiting other stars—has shown just how unearthly these planets can be. The first exoplanets to be discovered were orbiting a type of dead star called a pulsar. The next to be discovered were giant, Jupiter-sized (and larger) worlds orbiting extremely close to their stars. Planets around dead stars or super-Jupiters scorched by their suns are not other earths. Further discoveries in exoplanets have shown us that, contrary to Bruno's and Fontenelle's idea of a homogeneous Universe full of other suns and other earths, the Universe is full of a diversity of worlds we never imagined—planets with orbits like comets, planets dark as coal. In fact, half the planets discovered are of a size that does not even exist in our Solar System. In our Solar System, there are planets the size of Earth and Venus and smaller, and planets the size of Uranus and Neptune and larger (see Figure 1.2) but nothing in between. Indeed, astronomers once spoke as though such "in-between" planets could not exist.[4] Yet half the exoplanets discovered are in fact of these "in-between" sizes.

---

[4] For example, in 1991 David Black, then Director of the Lunar and Planetary Institute in Houston, wrote that "Giant planets probably form only in cool, distant parts of the solar nebula where water condenses into ice, providing abundant material for planetary growth. Planets in hotter regions closer to the forming star accumulate from less common silicon and iron compounds; they end up as smaller, rocky bodies like the earth." (See Black, "Worlds around Other Stars," 78.) Therefore, there would be either giant planets (far from a star) or Earth-type rocky planets (closer to a star). Mechanisms for the formation of "in-between" planets, now called "super-Earths" and "sub-Neptunes," were first proposed in 2008, before almost any such planets were known (Clement et al., "Formation of Terrestrial Planets," 3, 41–42).

Today's science simply adds to what was becoming clear at the turn of the twentieth century—that the Plurality of Worlds model of the Universe was in danger of collapsing. The turn of that century was then the second junction point at which that idea of the Universe arguably should have begun to wither—and then continued to wither because of the Big Bang model and of exoplanets just outlined.

It is worth pausing here to reiterate that science since the nineteenth century has been indicating that the Universe is not *full* of other earths and not *full* of life—but it has *not* been indicating that there exist no other earths, no other life, and no other intelligent life. The Universe is immensely large, and this fact has given rise to what we'll call the Big Numbers Maneuver, which unfolds as follows: Be as stringent as you like about what would make a planet count as an "earth":

- Insist the planet be of the right size, in the habitable zone of the right kind of star, in the right region of a galaxy.
- Insist that the planet have an oddly large moon, like Earth's Moon, to stabilize it and generate tides.
- Insist that it be in an oddly quiet planetary system, like ours.
- Recognize the difficulty of getting the right amount of water, such as Earth has, on what was once a molten-hot globe—too little and you have a desert planet; too much and it is a water world.
- Assume that the formation of life is an exceedingly rare event, for otherwise we would see it happening. Assume that even if simple life does form, the subsequent formation of *complex* life is then likewise rare, since current science suggests that the formation of complex life on Earth was the result of an incredibly unlikely merger of simpler life forms—a merger that happened *once* in Earth's history.[5]

Even given the stringency of these conditions, there are so many galaxies, stars, and planets that you are still going to end up with some other earths in this Universe. According to this Big Numbers Maneuver, there are bound to be other pale blue dots in the hostile dark. And yet they just might be very, very far away—*so* far away that chances are exceedingly small that there could ever be any moments of mutual discovery.

The theoretical, possible existence of *some* other earths does not change the fact that the start of the twentieth century appears as a point at which

<hr>

[5] Lane, *The Vital Question*.

science plausibly could have, and arguably should have, turned off the road to Pluralism and away from the idea of a Universe *full* of other Earths. *Yet it did not.* No further discussion of science is required to demonstrate this refusal. We can merely look at what has been popular entertainment since the turn of the twentieth century and see in the endless iterations of *Star Wars* and its kin evidence that the Pluralistic Universe did not wither at all. It flourished.

In summary, science has twice been in a position to question decisively the idea of a pluralistic Universe. But twice that Universe has endured in our imaginations anyway. Our fascination with other earths persists not because of scientific discoveries but because the idea is so marvelous, "so universally interesting."

And beyond the sheer excitement of imagining ETI, the prospect of Plurality places us among people ready, in the words of Ralph Cudworth in 1678, to "transcend those narrow limits, which vulgar opinion and Imagination" set upon the works of creation.[6] And of course it involves "going boldly," like William Shatner. Accounts of Plurality consistently fall back on this early modernist rhetoric of self-congratulation: In the late twentieth century, the massively scholarly work of Karl S. Guthke offered readers exhilarating prospects of "the last frontier." Such prospects promise an alternative to "the comforting notion of man's uniqueness in the universe"; and of course in the train of Copernicus, heliocentrism implied "that the uniqueness of the Earth, hitherto taken for granted, was an illusion."[7] Why would we ever admit to seeking comfort amid illusions and imaginations instead of daring—as *scientists*, no less—to explore other worlds? Perhaps this is why we have never turned off the road to Pluralism, the road to the Universe of *Star Wars* and kin, at one of those junctions offered by scientific evidence.

Of course, if we had turned off, we might not be searching for extraterrestrial intelligent life now, and we might not have been so certain we'd *found* it a century ago.

[6] Cudworth, *The True Intellectual System of the Universe*, 4:179.
[7] Guthke, *The Last Frontier*, 28, 43.

15

Searching for Intelligent Extraterrestrials

Right around the time when Agnes Clerke was recording that stars are not all suns, intelligent life was purportedly discovered on Mars, life that we might communicate with by means of radio signals. In December 1907, *The Wall Street Journal* gushed that "[t]he most extraordinary development [of 1907] has been the proof afforded by the astronomical observations of the year that conscious, intelligent life exists upon the planet Mars." It went on: "There could be no more wonderful achievement than this, to establish the fact of life upon another planet." Rapid developments in technology—the previous decade had witnessed Guglielmo Marconi's tremendous contributions in wireless telegraphy, or radio—would no doubt lead to establishing "some sort of communication with the people of Mars."[1] Over the following decade, the *New York Times* also ran numerous articles on Mars and the life there, including a lengthy piece about how the Martians had built two new planet-spanning canals in just a couple of years (see Figure 15.1).

To construct such long canals in such a short time, the Martians would have had to possess incredibly advanced technology. It was taking Earthlings a decade just to cut 50 miles of canal through the Isthmus of Panama. The Martians and their canals (which were supposed to irrigate their near-desert world with meltwater from the planet's polar ice caps) had been hypothesized by the American astronomer Percival Lowell (1855–1916) and others to explain a network of lines seen on Mars by various observers. In 1907 Lowell's team, having set up an observing station under the exceptionally clear skies of the Chilean desert, obtained high-quality (for the time) photos of Mars, which the *Wall Street Journal* apparently believed to prove the case (see Figure 15.2). In 1924 an effort was made to detect radio signals from Mars.

Needless to say, no signals were detected. Although Mars does have polar caps made of dry ice, not water ice, it is a desert world. The lines that gave rise to all the excitement were a result of optics, of the limitations of the eye and

---

[1] "Review and Outlook—Mars."

A Universe of Earths. Dennis Danielson and Christopher M. Graney, Oxford University Press.
© Oxford University Press (2025). DOI: 10.1093/9780197803547.003.0015

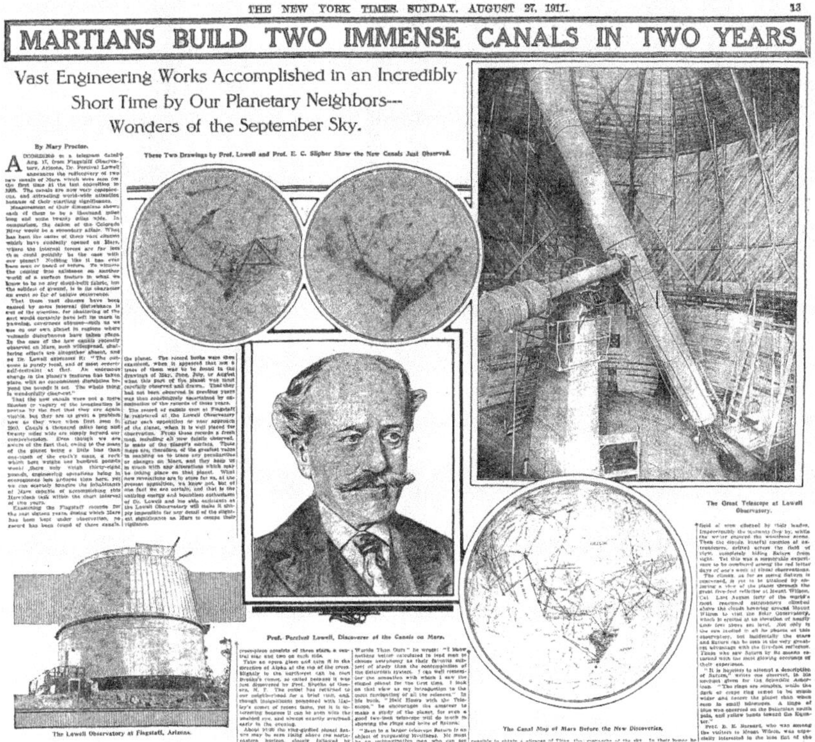

Figure 15.1 From *The New York Times*, August 27, 1911, on the activity of the inhabitants of Mars. The man at center is Percival Lowell.

of the telescopes and cameras of the time. The Martians were a product of these, too, of course—but also of the fact that for well over a century people had assumed and imagined that life existed elsewhere (perhaps everywhere) in the Universe. Plurality of Worlds!

Not all scientists assumed that such was the case. At least a few understood that for some time science had considered those assumptions to be disputable. In 1904, Alfred R. Wallace (1823–1913), co-founder with Charles Darwin of the theory of evolution by natural selection, published *Man's Place in the Universe* in which he urged the diversity of the planets and the uniqueness of Earth. "Not only is [Earth] situated at that distance from the sun which, through solar heat alone, allows water to remain in the liquid state over almost the whole of its surface, but it possesses numerous characteristics which secure a very equable temperature, and which have secured to it very nearly the same temperature during those enormous geological

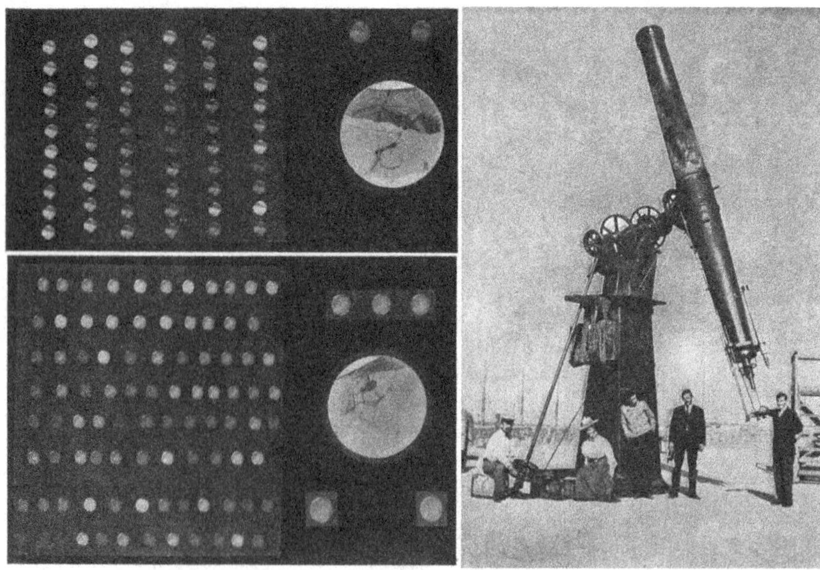

Figure 15.2 Photographs of Mars (small, left) with a drawing (large, center) of the same area of Mars as shown in the photos, published by Percival Lowell in the December 1907 issue of *The Century Magazine* (the quality of this reproduction is not optimal). Lowell wrote that "These little lines are the 'canals' which for their strange directness and yet stranger articulation were for long denied existence, and even now find a world slow to credit the story they have to tell. For it wounds man's dignity to believe it. But to the camera no evasion of the fact avails. . . . That life is there is founded on no assumption, but on massed evidence that is conclusive, and the reader should realize that opposition to the idea that we now have proof of life on Mars is not based on reason, but on emotion." Right: *Century Magazine* photograph of the expedition to the Chilean desert, where the skies are especially clear, to obtain the Mars photographs.

periods in which terrestrial life has existed. . . . [N]o other planet possesses these characteristics now, and it is almost equally certain that they never have possessed them in the past, and never will possess them in the future."[2] But which is more appealing, more (again to echo Brewster) "universally interesting"—Lowell's Martians or Wallace's lifeless Solar System?

Those Martians and their descendants still appeal and are still with us, in a sense. They had an enormous impact on popular culture, largely thanks to

[2] Wallace, *Man's Place in the Universe*, 273–74.

the British author H. G. Wells, whose fertile imagination extrapolated upon the "discovery" of Martian canals and imagined the advanced inhabitants of that dry planet attempting a conquest of resource-rich Earth. In his *War of the Worlds* (1898), the Martians came to Britain to do to the British what the British had allegedly done to other civilizations. Giant Martian three-legged walking fighting vehicles armed with heat rays devastated the English countryside, while the artillery of the British army and the ironclad battleships of the Royal Navy fought back with only sporadic success. Wells's book spawned an entire genre of popular science fiction and (later) film featuring "invading aliens." These included Orson Welles's 1938 *War of the Worlds* radio program, the 1953 *War of the Worlds* movie, the 1996 movie *Independence Day* and its 2016 sequel, plus yet another movie version of *War of the Worlds* in 2005, to name a few.

Of course, advances in science relegated Martians to pure fiction; Wallace was right, as was Whewell before him. In 1961, as the Space Age was dawning, the historian of astronomy Antonie Pannekoek recognized what arguably should have been recognized by Agnes Clerke's time:

> The dream of a plurality of inhabited worlds, the dream of other men living on neighboring kindred globes, is over. It is true that as yet we know nothing of possibilities in inaccessibly distant stellar systems; but as far as our own solar system is concerned, no other mankind exists than on earth.[3]

The passing of this dream has understandably had its impact on popular storytelling, since invaders from Mars have much less imaginative appeal once science has shown decisively that the planet in question lacks Martians. Whereas the earlier versions of the *War of the Worlds* story did feature Martian invaders, later versions featured invaders from those inaccessibly distant stellar systems where we might still imagine the "descendants" of Wells's Martians residing.

Ah, but might developments in communication technology allow us to establish some sort of contact with beings in those systems? Shortly before Pannekoek wrote that "the dream" was over, two scientists—Giuseppe Cocconi and Philip Morrison—wrote a paper that would bring that dream promptly back to life. They proposed searching for evidence of Pannekoek's other humankinds in distant stellar systems not by means of a telescope, such

---

[3] Pannekoek, *A History of Astronomy*, 393.

as Lowell employed in his search for evidence of alien life, but by means of *radio* antennas and receivers.

In 1959, these Cornell scientists published in the journal *Nature* their article "Searching for Interstellar Communications." In it they suggested that advanced extraterrestrial civilizations may be broadcasting radio signals that we are intended to hear. "No theories yet exist which enable a reliable estimate of the probabilities" regarding planets with intelligent life orbiting other stars, they wrote, but Earth at least is home to "a society recently capable of considerable scientific investigation." Maybe some such societies might even endure over long periods of time. "It follows, then, that near some star rather like the Sun there are civilizations with scientific interests and with technical possibilities much greater than those now available to us." They might be putting out radio signals that we could detect, especially since our own communication technology had improved rapidly during and since World War II.

Cocconi and Morrison argued that the cutting-edge technology of 1959 could pick up a signal "from any suitable star within some tens of light-years." The nearest star, Alpha Centauri, is four light-years distant, and Sirius, the brightest star in the night, is less than nine. Acknowledging possible resistance to their proposal, they wrote:

> The reader may seek to consign these speculations wholly to the domain of science-fiction. We submit, rather, that the foregoing line of argument demonstrates that the presence of interstellar signals is entirely consistent with all we now know, and that if signals are present the means of detecting them is now at hand. Few will deny the profound importance, practical and philosophical, which the detection of interstellar communications would have. We therefore feel that a discriminating search for signals deserves a considerable effort. The probability of success is difficult to estimate; but if we never search the chance of success is zero.[4]

Cocconi and Morrison assumed that extraterrestrial societies would be deliberately sending signals toward our Sun in hopes of establishing communication. However, *deliberate* efforts at interstellar communication are more than is needed for detecting extraterrestrial civilizations. As you read

---

[4] Cocconi and Morrison, "Searching for Interstellar Communications," 844–46.

this statement, all over the world radio and television stations are broadcasting news and entertainment; air traffic control and weather radar stations are scanning the skies; people are talking on cell phones. All these devices broadcast *electromagnetic waves* (which include radio, but also the infrared that we feel as heat, light, the ultra-violet that causes sunburns, the x-rays that are used in medicine, etc.). Much of what we broadcast simply travels out into space, without our intending it, moving at the speed of light. As these waves move away from their source, they become weaker in intensity, but they do not disappear.

Powerful radio stations have been broadcasting on Earth since the 1920s—for roughly a century at the time of this writing. The signals from those stations have traveled a hundred light-years out into space and accordingly have passed by a lot of stars. A hundred light-years is a tiny distance compared to the dimensions of the Galaxy as a whole, but there are still many stars within that distance of Earth. If any of those stars are orbited by a planet inhabited by extraterrestrials who have a very sensitive radio receiver, they might have picked up our signals. They might know that we are here. Likewise, if there are any nearby extraterrestrials broadcasting electromagnetic waves—putting out their own radio and television programs; operating their own radar stations; or as Cocconi and Morrison suggested, deliberately sending signals into space for communication—then we might detect their signals. All we would need is a sensitive enough receiver tuned to the right wavelength.

Cocconi and Morrison were not the only ones thinking about interstellar communication. Over a period of several months starting in April 1960, a team led by the American radio astronomer Frank Drake, whom we met at the start of this book, used a sensitive radio receiver at the new National Radio Astronomy Observatory in Green Bank, West Virginia, to search for extraterrestrial signals coming from two nearby, sun-like stars: Tau Ceti and Epsilon Eridani (see Figure 15.3). The project was the first scientific effort to detect alien civilizations among the stars. In 1979 Drake described the search, which was conducted using the receiver, a chart recorder that measured signal strength, a tape recorder, and at times a loudspeaker:

Whenever you search for extraterrestrial intelligent radio signals, you always feel at the beginning that the signal may pop up right away. And so the telescope operators and I spent a breathless morning peering at the wiggling pen on the chart recorder, thinking that every time the pen started

Figure 15.3 The radio telescope used by Frank Drake in 1960.
Credit: National Radio Astronomy Observatory (NRAO)

to deflect up that this was IT. Only to see the pen go down again, obeying the universal law of gaussian noise statistics. And so it went until noon, when Tau Ceti set in the west.

Then we turned the telescope to point at our second subject, the solar-type star Epsilon Eridani. . . . Again we pointed the telescope at the star, and set up the recorders. We had also added a loud speaker so that we could hear the receiver output. Again we started the chart and tape recorders, and settled back for more of what had already become routine.[5]

---

[5] Drake, "A Reminiscence of Project Ozma."

These sorts of searches came to be known as SETI—search for extraterrestrial intelligence. Since Drake's 1960 search, there have been many SETI searches, each attempting to listen to as many stars, on as many frequencies, as possible. The listening and monitoring are typically done by computer, not manually as in Drake's 1960 search. There was even a SETI program in which the average person could participate: "SETI@home." This creative program allowed people everywhere to download radio telescope data to their computers, which then processed the recordings, listening and monitoring while the computers were idle.

The creativity of SETI researchers has extended well beyond SETI@home to incorporate approaches to SETI quite different from the radio telescope work of the days of Drake, Cocconi, and Morrison. For example, there have been searches for signs of extraterrestrial intelligence (ETI) in other galaxies. The Universe is sufficiently old for extraterrestrial civilizations to exist that have possessed advanced technology for tens of thousands of years or more. Human civilization, by comparison, has possessed such technology for at most a couple of centuries. Such ancient, hyper-advanced civilizations might have an impact on their environments that is detectable even at great distances. For example, hyper-advanced extraterrestrials might build a giant structure around their home star for the purpose of collecting its energy or providing living space for a vast population. Such a "megastructure" would alter the energy output of the star versus the natural stellar output in a manner visible to our telescopes, and astronomers have sifted the data we have gathered on stars to search for such megastructures. A civilization acting on a galactic scale, or even over a portion of a galaxy, would likewise be detectable to our instruments, and astronomers have surveyed large numbers of galaxies to look for anomalies that might reveal such civilizations.[6]

A very different example of SETI creativity is the search within our Solar System for signs of extraterrestrial activity. Given the long spans of time over which extraterrestrial civilizations could have arisen, flourished, and declined, it is possible that our Solar System has been visited by extraterrestrials who left artifacts we might detect. If visiting extraterrestrials had built a research station on Earth 100,000 years ago, then all signs of it would likely

---

[6] Annis, "Placing a Limit on Star-fed Kardashev Type III Civilisations"; Billings, "Alien Super-civilizations Absent from 100,000 Nearby Galaxies"; Griffith et al., "The Ĝ Infrared Search for Extraterrestrial Civilizations with Large Energy Supplies. III. The Reddest Extended Sources in *Wise*."

have been destroyed by the same forces of water, wind, and biology that devour our own structures over time. But if such a station had been built on the Moon, where there is no water, wind, or biology, then remnants might still be there. We might train an artificial intelligence to scan vast numbers of Moon photos to hunt for those remnants.[7]

Despite these varied and creative approaches, and continuing radio searches, SETI has come up empty-handed. The fact that we can look at nearby galaxies and not see obvious evidence of a hyper-advanced civilization sprawling across one of those vast disks of stars does *not* rule out the possibility that extraterrestrial life exists. But the ongoing failure of SETI over more than half a century, despite much improvement in our instruments and methods of detection, is telling. The longer that SETI searches go on without detecting alien civilizations, the more probable it is that the Universe is not heavily populated with detectable advanced alien societies.

We might liken SETI searches to our reaching into an enormous, stadium-sized vat of M&Ms hoping to retrieve a golden one. (If you're in the UK, please feel free to substitute Smarties for M&Ms!) We reach in and grab a handful—and find no golds. Does that prove there are no golds anywhere in the vat? No. We try again—and again we get no golds. And again—and again no golds. After twenty attempts (or fifty, or a hundred), we still have found no golden M&Ms. The vat is still full of M&Ms. Have we proven that there are *no* gold ones? No. But, we have learned something about the golds, have we not? Were they abundant, they would be easier to find.

The shrinking probability of establishing communication with extraterrestrials is recognized from time to time. Even a quarter of a century ago enough SETI searches had been conducted, with enough thoroughness, that the prominent journal *Scientific American* published an article arguing that it was possible to determine with a fair degree of confidence that within 100 light-years of Earth there are no Earth-level civilizations deliberately broadcasting signals toward us, that within 1000 light-years of Earth there are no civilizations significantly more advanced than Earth, and that any hyper-advanced civilizations are confined to galaxies very far, far away (not in our Milky Way, or the Milky Way's neighboring galaxies).[8] A few years

---

[7] Adam Frank, *The Little Book of Aliens*, 148–50.
[8] LePage, "Where They Could Hide."

later, the director of the SETI program at Berkeley argued that certain kinds of hyper-advanced civilizations could be safely ruled out because "just from the astronomical data on file we would have discovered such a civilization by accident."[9]

SETI pioneers like Frank Drake and Carl Sagan hypothesized that our Galaxy might be full of advanced extraterrestrial civilizations whose radio signals we could detect. In the early days of SETI, Drake clearly thought he might pick up a signal right away. Obviously, that has not worked out. Even the idea of searching the Moon suggests limits—if those visiting ETIs of 100,000 years ago had built a sprawling research complex, with large structures for housing and transportation, we probably would have discovered some of them by accident as NASA made precision surveys of the Moon in preparation for the Apollo lunar landings. We are thus left with having our AI search the Moon for signs of only the sorts of ETIs who leave *very small* messes behind.

Such skeptical thoughts regarding ETI are not new. Enrico Fermi (1901–1954) reportedly responded to enthusiastic talk of intelligent life on other worlds by simply asking, "Where are they?" For if the universe teemed with ETIs in the ways that Plurality of Worlds enthusiasts imagined, then would we not already have seen evidence of them just by accident? The "Fermi Paradox" has long troubled SETI research and still does.

Of course, there are the Hollywood-style claims that SETI has succeeded or that we already have evidence of ETI in the Solar System: "The pyramids!"; "The face on Mars!"; "UFOs!"; "Area 51!" Drake noted how early in his SETI work rumors erupted of a big cover-up. A signal was detected that caused excitement among the team, but the source of the signal could not be determined before it disappeared.

> To our chagrin, one of our employees called up a friend in Ohio and told him about the signal. The word was passed to a newspaper reporter friend, and suddenly we were deluged with inquiries about the mysterious signal:
>
> > 'Had we really detected another civilization?'
> > "No."
> > 'But you have received a strong signal with your equipment?'
> > 'We can't comment on that.'

[9] Dan Werthimer, director of the SETI program at the University of California at Berkeley, quoted in Dorminey, "The Search for Astroengineers," 32.

And so, aha, we were hiding something. To this day many people believe falsely that we received signals from another world, and that some fiendish government agency has required us to keep this a deep dark secret.

Thus, the very first SETI search spawned rumors of a Grand Conspiracy that was seen as trying to cover up the fact that contact had actually been made with aliens. The signal later returned and was determined, by means of a second antenna, to be of earthly origin (at least that is what "they" want you to believe!).[10]

But more serious thought suggests that SETI simply exemplifies what history tells us: Hope for its quick success, such as Frank Drake harbored in 1960, or that the *Wall Street Journal* had demonstrated in 1907, was premised on a vision of a Bruno-esque Plurality of Worlds Universe in which other earths abound and in which therefore advanced extraterrestrial societies might be common enough to be readily detected. That view of the Universe should arguably (given the work of Kepler) never have become so well established, and even *if* established, it should have weakened significantly under the sorts of evidence for diversity in the Universe brought forward by William Whewell and Agnes Clerke in the nineteenth century. What is consistent with even the most creative SETI searches coming up empty-handed across six decades is a Universe in which Earth is joined in the dance of the stars by, yes, many worlds, but exceedingly few like it—a diverse Universe in which few among the stars are suns and perhaps fewer among the planets are earths; in which life is not a ready product of matter; and in which complex life might be the result of extremely improbable events.

[10] Frank Drake, "A Reminiscence of Project Ozma."

16

"An Awful Waste of Space"

Johannes Kepler thought the Moon had gravity. His argument—that telescopically observed lunar lakes *demonstrated* the Moon's gravity—turned out to be wrong. It was not unreasonable, however, and *more* reasonable than his own and John Wilkins's extrapolations of telescopic evidence in support of belief in an advanced lunar civilization, and also more scientific than later uses of heliocentrism to justify the image of Louis XIV as the Sun King. What Kepler would have called a lake we today (in Latin) still call *mare*, or a sea. When the Apollo 11 lander touched down on the lunar surface, Neil Armstrong reported, "Houston, Tranquility Base here, the Eagle has landed." *Mare Tranquilitatis*, the Sea of Tranquility, so dubbed by Giovanni Battista Riccioli, whom we met in Chapter 8, is a dark, basaltic plain, not a lake of water. Yet, clearly Kepler was not the only one who thought of lunar water in the early days of the telescope. His ideas about lunar water were wrong, but his intuition about lunar gravity was right.

Kepler also intuited that both our Solar System and our Earth were unusual—indeed unique. "Our little ball, the little cottage of us all, which we call the Earth," was a noble world that "may easily scorn the rest of the bulk." And then there were "these fine bits of dust called human beings . . . in whom is the image of God; who are, in a certain way, lords of the whole bulk." Kepler was happy to imagine other beings on the Moon and on Jupiter, but he understood that telescopic observations showed that, contrary to Bruno, stars were *not* other suns with other earths, and Earth was *not* just one of a vast swarm of similar worlds.

Our argument in this book has been that these Keplerian intuitions were also largely sound ones. Earth may have joined the dance of the stars, but it is not just one of a vast swarm of earths. As with his inference of lunar gravity, much of the science Kepler relied on has been superseded, but his intuition seems to have held.

Kepler's intuition ran counter to the Great Copernican Cliché asserting that the progress of science demoted the Earth, making it nothing special

A Universe of Earths. Dennis Danielson and Christopher M. Graney, Oxford University Press.
© Oxford University Press (2025). DOI: 10.1093/9780197803547.003.0016

(just another rock orbiting just another star). We've argued that Copernicus did *not* demote the Earth; instead, he raised it out of the downstairs of the two-storey Aristotelian Universe, up out of the sump, and made it a star, a celestial body. By inference, celestial bodies might then also be earths. But the sheer appeal of the idea of other earths caused us human beings to run ahead of the science, despite Kepler's protestations. Tossing aside not only the two-storey Universe but also good scientific evidence, our imaginations filled the Universe with earths, and those earths with civilizations—we dreamed of a Plurality of Worlds. Then, when our technology rendered us able to hunt for those beings with whom our dreams had filled the Universe, we commenced searching. Yet we have so far found no evidence to suggest that our dream of other earths is anything more than a dream: Despite more than six decades of SETI, starting with Frank Drake's 1960 search for radio signals from another star, none has been detected.

In Chapter 1 we mentioned the much-loved truism that absence of evidence is not evidence of absence (of intelligent extraterrestrials). Another favorite truism is some version of Cocconi and Morrison's "if we never search the chance of success is zero." And nothing we have written here should be construed as a case for not searching, not looking for evidence. Yet our emphasis is precisely on *evidence*. The case for SETI—and for Plurality as a whole—will remain conspicuously incomplete if the kind of evidence sought or offered fails to reach beyond "analogical reasoning," or the familiar Big Numbers Maneuver that drives to a particular Pluralistic conclusion, which underpinned the entire Pluralistic enterprise for centuries and continues to be used today.

A typical instance of the Big Numbers Maneuver appears in the movie *Contact* (1997), based on Carl Sagan's novel of the same name (1985). The story's main character, Dr. Eleanor "Ellie" Arroway, is based on the real-life Dr. Jill Tarter (still a central figure in SETI) and played by Jodie Foster. Ellie's romantic interest in the film is Palmer Joss, played by Matthew McConaughey. The following dialogue takes place as they together gaze up into the night sky while sitting by the Arecibo radio telescope:

Ellie:
There are 400 billion stars out there, just in our Galaxy alone. If only one out of a million of those had planets, all right, and if just one out of a million of those had life, and if just one out of a million of those had intelligent life, there would be literally millions of civilizations out there.

Palmer:
Well if there wasn't, it'd be an awful waste of space.
Ellie:
(Pause) Amen.

Ellie is, in effect, walking Palmer through a simplified "Drake Equation" calculation (named for Drake, of course) to estimate how many civilizations might exist in our Galaxy or in the Universe. Ellie's Big Numbers Maneuver and Palmer's response reflect a centuries-old feature of the case for Pluralism. Ralph Cudworth wrote in 1678 that "it is not reasonable to think that all this immense vastness should lie waste, desert, and uninhabited, and have nothing in it that could praise the Creator thereof, save only this one small spot of earth."[1] Even the atheistic Ellie's reverent "Amen"—an affirmation usually reserved for statements of faith—oddly resonates with Cudworth.

Palmer's "waste of space" comment echoes sentiments we discussed in Chapter 5 by the sixteenth-century poet Marcellus Palingenius. It reflects a long-standing argument for the existence of extraterrestrials. What might the probative force of such a comment be? It appears to be tied to what A. O. Lovejoy dubbed "the Principle of Plenitude": the idea that a perfect creation must in some sense be "full," with no gaps.[2] From Aristotle right through to Descartes in the seventeenth century, it was common wisdom that "Nature abhors a vacuum." One of the most concise expressions of this principle was put into verse by Alexander Pope, in his *Essay on Man* (1733):

> Of systems possible, if 'tis confest
> That Wisdom infinite must form the best,
> Where all must full or not coherent be,
> And all that rises, rise in due degree;
> Then, in the scale of reas'ning life, 'tis plain
> There must be somewhere, such a rank as man.

Are Palmer and Ellie embracing the ancient and medieval—indeed, the teleological and theological—roots of this way of thinking? Surely not, at least not knowingly. But *if* not, then it's hard to grasp what the "waste of space" appeal either presupposes or is intended to prove. Certainly from the time and work of Blaise Pascal (contemporary with Descartes) and others

---

[1] Cudworth, *The True Intellectual System of the Universe*, 4:180.
[2] Lovejoy, *The Great Chain of Being, passim.*

who empirically established that nature does *not* abhor a vacuum, modern science, including modern astronomy, has not found the idea of empty space offensive or objectionable.

Finally, Ellie's Drake Equation numbers actually imply a lot of wasted space: *400 billion stars out there*, in our Galaxy; *one in a million has planets* (she must mean Earth-like planets, since, as we have seen, astronomers have long assumed that stars as a norm have planets)—so 400 billion divided by a million is 400 thousand; *one in a million of those planets has life*—400 thousand divided by a million is 0.4; and *one in a million of those have intelligent life*—0.4 divided by a million is 0.0000004. That means four civilizations for every ten million galaxies like ours! To have "literally millions of civilizations out there" would require a Universe of tens of *trillions* of galaxies. Current estimates tend to top out around two trillion galaxies, so by Ellie's numbers there would not, in fact, be millions of civilizations out there, and the civilizations that did exist would be very widely scattered (one for every 2.5 million galaxies). Were they that scattered, the chances of detecting them would be nil, our radio telescopes should be put to better use, and there would indeed be an awful lot of "wasted space." The numbers have to be much better than Ellie's repeated "one in a million."

It is odd that Ellie's numbers appeared in a movie based on a book by Carl Sagan. At an international SETI conference in 1971, organized by Sagan and held in Byurakan (Soviet Union), the consensus reached by those attending was that a million advanced civilizations existed in our Galaxy, a number Sagan would urge on his own.[3] That consensus required far, far more optimistic numbers than Ellie's. Indeed, the Byurakan conference was far too optimistic. Let us borrow Ellie's maneuver here—*if just one out of a million* of those extraterrestrial civilizations in a Galaxy like ours were hyper-advanced, there would be a hyper-advanced civilization in every Galaxy, and as our increasingly powerful (since 1971) telescopes studied all those Galaxies, and our own Galaxy, surely we would have seen some sign of them, "just from the astronomical data on file." What Fermi said.

We are not arguing that our telescopes should be put to better use than SETI—a position argued in 1975 by astronomer Michael Hart. "An extensive search for radio messages from other civilizations is probably a waste of time and money," he wrote, based on evidence then available. "The idea that

---

[3] Dick, *Life on Other Worlds*, 215–17; Sagan, "Quest for Extraterrestrial Intelligence," 2–3: "When we do the arithmetic, the number that my colleagues and I come up with is around a million technical civilizations in our Galaxy alone."

thousands of advanced civilizations are scattered throughout the Galaxy is quite implausible," Hart asserted, warning against a search based on "wishful thinking."[4] (NASA historian Steven Dick called his argument "The Hart Attack.")[5] Certainly we can endorse Paul Davies's claims that "SETI is fundamentally an experimental and observational programme, not an exercise in philosophy and statistics," and that "An eerie silence is no reason to abandon the search."[6] But at the same time, we do need to acknowledge the silence (not so eerie when you know the history of Pluralism) and to state plainly, scientifically—repeating a favorite tautology of our own—that absence of evidence *is* absence of evidence. The Big Numbers Maneuver, analogies, and assumptions about what cosmic teleology requires do not constitute genuine evidence.

And yet, absence of evidence *is* evidence of absence at a certain level. To return to our own analogy from the last chapter, of the vat of M&Ms: the vat is huge; we will never hope to sift through even a tenth of its contents; but with every handful of candy that we pull out that produces no golds (that is, with every new telescopic observation made with our increasingly powerful telescopes that yields another negative SETI result), we accumulate evidence that the golds, while they might exist, are few and far between. And if things keep going as they have been, we will eventually decide to stop actively searching the vat for golds, especially when we begin to look at our history and to realize that what we have been assuming for centuries—that the vat is teeming with gold M&Ms—is without scientific basis.

The absence of scientific evidence for filling the Universe with populated earths has not prevented us human beings from indulging in a persistent sort of Pluralistic misanthropy—a misanthropy that seems absurd given its lack of scientific foundation. We authors are not disinterested parties to the conversation and the history we're pursuing in this book. As should be obvious to anyone who has read our account thus far, we have affection for history and literature and science, affection rooted, simply and elementally, in profound love for our home and native planet, for its inhabitants, and for getting their story right. So we are astonished—not only at the glories of the Earth and of the Cosmos that surrounds us, of which we humans are unarguably a part—but also, and *sadly*, at what we have called a "misanthropic undercurrent" found in discussions of other worlds. Recall Huygens's rhetorical query

---

4 Hart, "An Explanation for the Absence of Extraterrestrials on Earth," 134–35.
5 Dick, *Life on Other Worlds*, 219.
6 Davies, *The Eerie Silence*, p. 92.

(1698) "whether Nature has laid out all her cost and finery upon this small speck of dirt." Recall Kant's remarks about a louse reflecting, "in its way of living as well as *in its lack of worth* . . . the condition of most human beings." Of course, this undercurrent does not apply to every enthusiast for Pluralism: consider Frank Drake, whose manning of a suicide hotline indicates his deep appreciation for the worth of human life.

Moreover, the exhilaration accompanying the recognition of Earth's post-Copernican "star-status" certainly needn't preclude a probing human or planetary self-critique. It may be true, to adapt a phrase from Winston Churchill, that as a location and as a species we merit the epithet "humble"—with much to be humble about! Some astronomers speak of humans' "existential humility."[7] But such self-critique and the mandate that Earth or humanity should not be too cosmically "uppity" ought not to be based on outworn physical or metaphysical notions inherited (consciously or not) from a pre-Copernican two-storey Universe.

We've already noticed cases in which scientists and serious thinkers had trouble transcending the physics or metaphysics of a world view that had held sway for many centuries. Even a brilliant, enthusiastic Copernican convert like Thomas Digges couldn't shake off the notion that our Earth—while, yes, a planet—was nevertheless still to be thought of as a "*dark* star." In the century after Digges, despite growing acceptance of Copernicanism, at the popular level it may have seemed just too counterintuitive that Earth should now be considered ethereal, heavenly. As Edward, Viscount Conway, wrote in 1651: "In my opinion, that [opinion] of Copernicus, for the Earth a heavy dull gross body to move and the heaven and stars who are light to stand still is as if a prince should upon a festival day appoint all the old and fat men and women to dance and all the young men and women of sixteen and twenty to sit still."[8] It simply seemed contrary to what for hundreds of years had passed as common sense: It was just too hard to believe.

But "just too hard to believe" cannot be passed off as scientific evidence, even if scientists themselves may be as susceptible as anyone else to deeply entrenched habits of thought—a susceptibility evident in the debates set off by William Whewell's 1853 *Of the Plurality of Worlds*, discussed

---

[7] For example, Vanessa Bailey of NASA: See Dave Eggers, "The Searchers: Dave Eggers on NASA's Jet Propulsion Lab," *The Washington Post* (September 17, 2024), https://www.washingtonpost.com/opinions/interactive/2024/dave-eggers-jet-propulsion-laboratory-nasa-who-is-government

[8] Edward, Viscount Conway, letter to his daughter-in-law Anne Conway, July 22, 1651, in Nicholson, *Conway Letters*.

earlier. Whewell argued *on scientific grounds* against the idea of a Universe widely inhabited with intelligent life. But at the time, his argument was treated as incredible and *unorthodox*. The Plurality of Worlds was an article of faith. We have glimpsed how many thinkers affirmed that other planets in our Solar System had their inhabitants. Recall the comments of Whewell's prominent critic, David Brewster, that Whewell's arguments must be ascribed "only to some morbid condition of the mental powers, which feeds upon paradox and delights in doing violence to sentiments deeply cherished and to opinions universally believed." Neither "deeply cherished" nor "universally believed" remotely equates to "supported by scientific evidence." Yet Brewster's language reminds us again that in his time the Plurality of Worlds did amount to a kind of orthodoxy. The same was acknowledged by Rev. (later Cardinal) John Henry Newman in his *Grammar of Assent* (1870): "In the controversy about the Plurality of worlds, it has been considered . . . to be so necessary that the Creator should have filled with living beings the luminaries which we see in the sky, and the other cosmical bodies which we imagine there, that it almost amounts to a blasphemy to doubt it."[9]

We may eventually ask whether things have changed much over the past century and a half or so. But one contemporary of Whewell's and of Brewster's, Samuel Warren (1807–1877)—a lawyer and a novelist—employed his satirical pen to sum up the two principal characters on either side of the nineteenth-century debate, characters he dubbed "Star-Smasher" and "Star-Peopler," respectively. For Brewster, Warren suggests, "wherever there is matter there must be life," right out to and including the nebulae. So no wonder at Brewster's "alarm . . . and anger" when he beheld Whewell "go forth on this exterminating expedition through Infinitude! It was like a father gazing on the ruthless slaughter of his offspring. Planet after planet, satellite after satellite, star after star, sun after sun, . . . nebula after nebula, all disappeared before this sidereal Quixote!"[10]

Yet it's unclear how many contemporaries could look upon the Plurality debate with similar humor or amusement. A perhaps more typically dour response was that of the poet Alfred Lord Tennyson (1809–1892), whom Whewell had tutored at Cambridge. According to Hallam Tennyson, the poet's son and biographer, the senior Tennyson found Whewell's essay on Plurality "anything but a satisfactory book. It is incomprehensible that the

---

[9] Crowe, *The Extraterrestrial Life Debate, Antiquity to 1915*, 362.
[10] Crowe, *The Extraterrestrial Life Debate, Antiquity to 1915*, 366.

whole Universe was merely created for us who live in this third-rate planet of a third-rate sun."[11]

Again we may express amazement at the firmness of such a view held by elite members of a scientifically literate society—views supported by no scientific evidence whatsoever, yet pertaining to a matter Whewell had subjected to scientific scrutiny. And, let us recall, Whewell was right beyond his own imagination. The Solar System outside of Earth is far more lifeless than he imagined (he allowed for gelatinous creatures on Jupiter, after all). Being neither psychologists nor sociologists, we can offer but little insight into what might be the sources of, or motivations for, such gratuitous defamation of our planet, unless they be perhaps some unconscious affinity for the old two-storey Universe with its lowly Earth. Suffice for now, however, that we register our sad amazement at the disparagement—together with our charge, again, that it has no basis in scientific evidence. Surely Earth deserves better. And we acknowledge the psychological and affective attractions of Plurality—along with the fact, as has already become apparent, that shopworn orthodoxies can emerge where one might least expect them.

[11] Crowe, *The Extraterrestrial Life Debate, Antiquity to 1915*, 365.

17

"Inevitably Privileged to Some Extent"

In the middle of the twentieth century, astronomer Hermann Bondi invoked the authority of Copernicus to support the "steady-state" cosmology with its eternal Universe that he had been promoting with two Cambridge colleagues, Thomas Gold and panspermia advocate Fred Hoyle. Bondi coined the term *Copernican Principle* to sum up the idea that Earth "is not in a central, specially favoured position" in the Universe.[1] Since that time, the Copernican Principle (CP), or some version of it, has continued to be endorsed by scientists even though steady-state cosmology has largely been abandoned.

In 1973, the year of Copernicus's quincentenary, Stephen Hawking and George Ellis published *The Large Scale Structure of Space-Time* and enlisted the CP to serve Big Bang cosmology. The geometry of the expanding Universe is such that things appear the same in all directions no matter where the observer is located. Other writers associate the CP with a Principle of Mediocrity, which reiterates in a nutshell that Earth's location in the Universe is consistent with our point of observation *appearing* to be in the cosmic center—but only trivially so, because just any other location would likewise seem to offer observations made from a universal center point. An analogy would be different observers located on the surface of an expanding sphere: Each observer, regardless of location, sits at the center of his or her encircling visible horizon and is equidistant from all points on that horizon and from the farthest limit of the sphere's end point (and is getting farther as the sphere expands).

The danger, however, is that a valid scientific insight can get applied in nonscientific ways. Hawking and Ellis, rather than restricting themselves to an explication of cosmic geometry, openly admit into their account of the Copernican Principle what they call "an admixture of ideology":

Since the time of Copernicus we have been steadily demoted to a medium sized planet going round a medium sized star on the outer

---

[1] Bondi, *Cosmology*, 13.

A Universe of Earths. Dennis Danielson and Christopher M. Graney, Oxford University Press.
© Oxford University Press (2025). DOI: 10.1093/9780197803547.003.0017

edge of a fairly average galaxy, which is itself simply one of a local group of galaxies. Indeed we are now so democratic that we would not claim that our position in space is specially distinguished *in any way*.[2]

Karl S. Guthke declares that "The new cosmologic model . . . deprived the Earth of *any* special status." Guthke goes on to assert, "not least of the implications [of heliocentrism] was that the uniqueness of the Earth, hitherto taken for granted, was an illusion."[3] The Copernican Principle or the Principle of Mediocrity thus becomes a sort of extension of the Great Copernican Cliché we discussed in Chapter 7.

Earth's standpoint truly is "mediocre" as pertains to the large-scale geometry of space. But that does not justify an overgeneralized denial of earthly excellence—a claim that Earth is mediocre in the nontechnical, popular sense of "undistinguished" or "low quality." One thinks of the gratuitous misanthropic tendency exemplified by Tennyson's reference to "this third-rate planet" of ours. A sloppy extension of perfectly good scientific vocabulary can thus seem to entail the cavalier denigration of our Earth. That denigration is a shame, and not very logical, given that our most excellent Earth is rather first-rate, being the only planet in the Solar System on which we human beings, including the denigrators, can survive! Such denigration, moreover, presumes an abundance of other earths, a Plurality of Worlds. In that case, not only would things appear the same in all directions no matter where in the Universe an observer were located, but also there would be an observer at all those locations, making one place truly the same as another, and not distinguished "in any way."

However, the past half-century or so has seen increasing resistance to any treatment of Copernicus's legacy that includes blanket denials of human or terrestrial specialness. In a seminal letter to the journal *Nature* in 1961, R. H. Dicke pointed out that the very existence of physicists who study the size and age of the Universe creates a "selection effect" implying that the Universe must be of a certain size and age because "It is well known that carbon is required to make physicists."[4] And carbon presupposes a stage of cosmic evolution within a (Big Bang) Universe large enough and old enough that the chemistry that constitutes observers can form and be distributed.

---

[2] Hawking and Ellis, *The Large-Scale Structure of Space-Time*, 134 (emphasis added).
[3] Guthke, *Last Frontier*, 36, 43 (emphasis added).
[4] Dicke, "Dirac's Cosmology and Mach's Principle," 440.

Recall that in the evolving Universe of the Big Bang, the Universe is initially so hot that even atoms could not exist. Therefore, carbon must somehow form before carbon-based life, including physicists who study the age of the Universe, can exist. In an eternal, unchanging, Universe, by contrast—like the two-storey Universe of Aristotle or the Universe of steady-state cosmology—carbon is always there.

What Dicke said echoes in a sense what William Thomson said almost a century earlier when talking about the molten globe of the Earth: Certain times in the Universe's history allow for us Earth-bound observers to exist; others do not. Our current point in *time* within the Universe must be privileged in some sense, even if we presume our *location* within it is not. In the auspicious year 1973—Copernicus's quincentenary—Brandon Carter articulated this insight at a symposium in Krakow, Poland. As a counterbalance to any indiscriminate generalization of the Copernican Principle, Carter proposed the Anthropic Principle, famously asserting that "Although our situation is not necessarily *central*, it is inevitably privileged to some extent."[5]

Since the appearance of Carter's paper, a large literature has grown up around the Anthropic Principle, and we won't attempt a full exposition or defense of this idea here. It is, however, highly pertinent to questions about a Universe that contains physicists, astronomers, and other curious minds. Perhaps the fullest single discussion of the Anthropic Principle, that of John Barrow and Frank Tipler, opens with a direct reference to the CP and the issue of our own or Earth's specialness:

The expulsion of Man from his self-assumed position at the center of Nature owes much to the Copernican principle that we do not occupy a privileged position in the universe. This Copernican assumption would be regarded as axiomatic at the outset of most scientific investigations. However, like most generalizations it must be used with care. Although we do not regard our position in the universe to be central or special in every way, this does not mean that it cannot be special in *any* way.

Barrow and Tipler go on to show how our position in time might be special. As the Universe cools from the Big Bang, the simplest elements form: lots of hydrogen, a little helium, and almost nothing else. Producing the stuff

---

[5] Carter, "Large Number Coincidences and the Anthropic Principle in Cosmology," 291.

we're made of, the aforementioned carbon, and all the oxygen in all the water that is in us, takes time:

> In order to create the building blocks of life—carbon, nitrogen, oxygen, and phosphorus—the simple elements of hydrogen and helium which were synthesized in the primordial inferno of the Big Bang must be cooked . . . for a much longer time than is available in the early universe. The furnaces that are available are the interiors of stars. There, hydrogen and helium are burnt into the heavier life-supporting elements by exothermic nuclear reactions. When stars die, the resulting explosions which we see as supernovae, can disperse these elements through space and they become incorporated into planets and, ultimately, into ourselves. This stellar alchemy takes over ten billion years to complete. Hence, for there to be enough time to construct the constituents of living beings, the Universe must be at least ten billion years old and therefore, as a consequence of its expansion, at least ten billion light years in extent. We should not be surprised to observe that the Universe is so large. No astronomer could exist in one that was significantly smaller.[6]

In other words, science now tells us that the Universe must be huge in order for us to exist. What an interesting thing this is: It is often said that Copernicus minimized our planet's relative cosmological importance given that heliocentrism entailed an enormous expansion of large-scale cosmic dimensions. Of course, science has always showed Earth to be small; we have seen how Ptolemy showed that Earth was "as a point" in relation to the sphere of the fixed stars, and "a point" is arguably as small as can be. But under Copernicus, the entire *orbit* of Earth was reduced to a point within the Cosmos. This meant that Copernicus's Universe was many orders of magnitude larger than that of Ptolemy, and of course that Earth was therefore orders of magnitude *smaller* by comparison. Digges's idea that we are in a universe of stars that extends out infinitely (perhaps "indefinitely" is a more fitting term, given what we know today) made Earth smaller still, by comparison.

That diminution might indeed, at least superficially, deal a serious shock to terrestrial self-esteem by indicating how trivial and insignificant we are.

---

[6] Barrow and Tipler, *The Anthropic Cosmological Principle*, 1, 3.

There's much in our ordinary way of speaking that assumes "bigger is better." The term *great* is one that we use to denote not only size but also value. We describe an admirable person as "great" or "big-hearted" or "magnanimous," and a less admirable person as "small-minded" or "pusillanimous."

For many centuries the valuation of the large and the devaluation of the small have played an important role in cosmological discussions. In the sixth century, Boethius exclaimed:

> How limited this glory is! How frivolous and how contemptible! You have learned from astronomy, that this globe of earth is but as a point.... Deduct the space occupied by the seas and lakes, and the vast sandy regions which extreme heat and want of water render uninhabitable, [and] there remains but a very small proportion of the terrestrial sphere for the habitation of men. Enclosed then and locked up as you are, in an unperceivable point of a point, do you think of nothing but of blazing far and wide your name and reputation? What can there be great or pompous in a glory circumscribed in so narrow a circuit?[7]

Sagan wrote pretty much the same thing in the twentieth century regarding our "pale blue dot":

> The Earth is a very small stage in a vast cosmic arena. Think of the rivers of blood spilled by all those generals and emperors so that, in glory and triumph, they could become the momentary masters of a fraction of a dot.... Our posturings, our imagined self-importance, the delusion that we have some privileged position in the Universe, are challenged by this point of pale light. Our planet is a lonely speck in the great enveloping cosmic dark.[8]

Of course we have ways of countering this pattern of thought. Bigger is not always better or more important. You may have had the experience of watching a loved one board a plane, which then takes off and gradually recedes from your vision, disappearing like a speck in the distance. In this and in countless other similar experiences we emphatically do *not* conclude, "Aha, see then how trivial and insignificant my loved one really is." On the contrary, we search for metaphors to express how worth may be embodied in

---

[7] Boethius, *Consolation of Philosophy*, 67.
[8] Sagan, *Pale Blue Dot*, 6–7.

tiny things, how love is not something merely bound by space or physical magnitude, how a precious jewel may dwarf in value the kingdom that contains it. Kepler too, as we saw in Chapter 1, sought to value us tiny human beings, us "fine bits of dust," within the rapidly expanding horizons of his Copernican Universe. In the same way today, surely a meditation on our pale blue dot requires a balancing of perspectives. Yes, it imparts a healthy humility to realize how little we and our planet are, cosmically speaking. But it also fills us with awe to contemplate, if we can, the inestimable riches concentrated in that small, pale, blue point of reflected light.

And yet Barrow and Tipler show—scientifically—how in this evolving Universe to which Copernicus led us, our existence *implies* our smallness; if we are to exist, we must be as points within a Universe so large that light, which is so fast that it blazes from here to the Moon in under two seconds, would take billions of years to cross it. If the sole purpose—or even *a* purpose—of the Universe were to create creatures like us, thereby allowing, as Sagan said, "the cosmos to know itself," then the Universe would have to be that large. Surely Johannes Kepler would be turning cartwheels over this revelation of science: "these fine bits of dust called human beings; to whom the Creator has granted such, that . . . [they] are, in a certain way, lords of the whole bulk. . . . For he glories not in bulk, but ennobles those he willed to be small." So yes, cosmically speaking, in some highly significant respects size does matter, in a surprising way. Our own and Earth's presence in the immense Universe points to an awe-inspiring coherence, even if for now most of us are limited to simply wondering at the connections.

There is another crucial respect in which vast cosmic size matters a great deal: that of the widespread human longing to discover—or perhaps be discovered by—inhabitants of other worlds. The mind-boggling distances even within the Milky Way, which is one galaxy among hundreds of billions of others, present serious challenges to any hoped-for contact among Earth and extraterrestrial civilizations "out there," if such there be. As we noted in Chapter 1, Earth's "radio bubble"—the imaginary sphere formed by the farthest reaches of terrestrial radio broadcasts (which began about a century ago)—now extends to a diameter of only two hundred light-years, or approximately 0.002 (one fifth of one percent) of the galactic breadth. Sheer size renders even waves traveling at the speed of light as downright plodding, sluggish, and incapable of reaching the cosmic territories we wish we could explore.

In response to the discouragements to SETI entailed by the cosmic immensities together with the finite and inviolable universal speed limit—300,000 kilometers or 186,000 miles per second—Harvard astronomer Howard A. Smith has introduced what he calls the Misanthropic Principle. We have already had occasion to complain of some astronomers' misanthropic tendencies as, for example, in the apparent relish some take in the supposed cosmic demotion and diminishment of humankind. But Smith uses the word "misanthropic" in a quite different, much less negative sense as pertains to our place in the Universe.

Smith reviews some of the threads we have already seen running through the Great Copernican Cliché and through arguments for a Plurality of Worlds, such as what we've called the Big Numbers Maneuver. Scientists have typically, Smith says, assumed that in a Universe "as spacious and rich as ours" there are "more than enough stellar systems to overwhelm 'pessimistic' scenarios," and that "there should be many stars with Earth-like planets hosting life because there is no reason to think that suitable planets or life are extraordinary."[9] In this vein, Smith cites Harlow Shapley, late director of the Harvard College Observatory, who "wrote of 'intimations of man's inconsequentiality' in a vast cosmos and of 'our [firm] belief in the cosmos-wide occurrence of life.'"[10]

To such "other worlds" optimism, Smith gently applies the counterweight of skepticism and a repeated insistence on the limitations imposed by the speed of light. Yes, we are discovering exoplanets, although they are turning out to be more varied than anyone had expected. Finding a planet in a habitable zone—not too far from or too close to its star, not tidally locked, in a nearly circular orbit (like Earth's) about "a host star stable in size, age, and radiative output for the billions of years needed to nurture intelligent life," perhaps (again like Earth) with plate tectonics, an ozone layer, a protective magnetic field, and a large moon to stabilize its rotation[11]—is highly challenging. And besides, "habitable by no means implies inhabited. . . . No wonder there are no signals, not even faint traces, despite decades of looking."[12]

---

[9] Smith, "Alone in the Universe," 499.

[10] Smith cites Harlow Shapley, *The View from a Distant Star*, 3, 77.

[11] For a much fuller list of "rare earth factors," see Ward and Brownlee, *Rare Earth*, xxvii–xxviii. See also Scharf's review of the unusualness of our Solar System, in *The Copernicus Complex*, 116–19.

[12] Smith, "Alone in the Universe," 502.

Smith writes that the question most often posed to him in this context is whether future knowledge and advancing technique might not offer a breakthrough permitting faster-than-light travel—"warp drive" or "hyperspace," to borrow from the world of science fiction. His reply is this:

> If life were common and if superluminal travel or communication were possible, then we face an insuperable contradiction: the billions of intelligent species in the universe (and we are surely one of the newest and least technologically advanced) should have already used this super-technology to visit us. That NONE have done so surely means that the basics of relativity as we know it will remain inviolable—leaving us alone. Alternatively, if relativity can some day be overcome, then there cannot be very many civilizations out there, and we are alone.[13]

Again, we hear the echo of Enrico Fermi's question: Where are they?

But the Misanthropic Principle, Smith emphasizes, is not as gloomy or pejorative as the adjective usually implies. On the contrary, it suggests that such are the cosmic impediments to the development and flourishing of intelligent life that Earth and its inhabitants may truly *not* be ordinary or "mediocre" in the popular dismissive sense. Instead, "Even though Earth is not at the center of the universe, its luxuriant environment could nonetheless make it a rare oasis. Perhaps we can appreciate that humanity, too, could be unusual, even special and not mediocre, at least as far as we are likely to know for a very long time."[14]

Note that Smith does not deny the existence of extraterrestrial intelligence (ETI). How could such a denial ever be scientifically supported? His claim is, rather, that—again given the known vastness of cosmic distances and the limits imposed by the speed of light upon both travel and communication—"For all intents and purposes, we and our descendants for at least 100 generations are very likely living in solitude." "ETI is almost certainly nowhere nearby, and its possible existence elsewhere is for all practical purposes unknowable to us (at least for a very long time, if not forever). Speculation about its nature is consequently specious and irrelevant. Very likely we are alone."[15]

Furthermore, such a conclusion need not be a counsel of despair! Yes, it is sobering indeed. But it includes a recognition of

---

[13] Smith, "Alone in the Universe," 502.
[14] Smith, "Questioning Copernican Mediocrity," 238.
[15] Smith, "Alone in the Universe," 512.

evidence that humanity is precious. The Earth, even if it is not unique, is for all intents and purposes a special place. The implication of the Anthropic Principle is that it might matter. The implication of the Misanthropic Principle is that we have to care for our planet and one another by ourselves—without help from alien insights or technologies. Modern science has prompted this re-evaluation, but addressing it will require the best of all our human abilities.[16]

Of course, the argument we are making in this book is that science has been prompting this reevaluation since Kepler argued star sizes against Bruno.

Yet, Kepler might caution Smith not to put too much confidence in modern science or in the two conclusions he draws from it: that we are "for all practical purposes" alone, and that speculation about life on other worlds is specious and irrelevant. Kepler drew his conclusions about stars being giant, dim bodies that were nothing like the Sun based on rock-solid, reproducible observations and impeccable logic—but it turned out that there was something about those observations that neither Kepler nor any other astronomer of his time knew, and his conclusions turned out to be wrong.

As we have pointed out several times, we don't insist that we are necessarily alone, and we don't consider it specious or irrelevant to think about, and even to search for, life on other worlds. But we do think it is important to know the history of science and especially how it pertains to the idea of life on other worlds. We do consider it important to understand that this thinking and searching are motivated by a human dream (as Antonie Pannekoek termed it), not by science. We human beings search for other earths and life on them, "for new life and new civilizations," not because of science, but rather in the hope that our science may be wrong or incomplete. That does not mean, however, that Smith is wrong about the pressing need for a reevaluation.

[16] Smith, "Questioning Copernican Mediocrity," 239.

18

A Precious Planet Earth

Copernicus taught us that our Earth is a star, a planet. And for almost five centuries now, science has been telling us that Earth indeed *is* a star in that other sense: *the star*, the garden spot of the Solar System and probably a garden spot of the Galaxy. Meanwhile, we human beings have been "dream[ing] of other men living on . . . kindred globes," as Antonie Pannekoek put it. We have been imagining a universe full of garden spots, ignoring the science as best we can while thinking of that dream as a product of science.

What does it mean if the idea of a Universe full of other earths—the infinite Cosmos of Giordano Bruno, the Plurality of Worlds Universe—has never been supported by science? It means a lot and it matters a lot for us, for science, and for planet Earth.

It matters because the idea of abundant other earths, and the scientific plausibility of that idea, are important components of our popular culture and popular imagination, as reflected in *Star Wars*, *Star Trek*, the Marvel Cinematic Universe—not to mention all the other pop culture materials featuring "aliens" (extraterrestrials) that are not part of these huge franchises (Steven Spielberg's *E.T. the Extra-Terrestrial*, for example). Consider the differences between Mr. Spock of *Star Trek* and another popular character who is a nonhuman intelligent life form: Ariel of Disney's *The Little Mermaid*—between what we call "*science* fiction" and what we call "fantasy." Where does it leave popular culture and popular imagination if abundant other earths are a fantasy with no more scientific support than the undersea kingdom of Ariel's father, King Triton?

There is more at stake here than simply whether the enormous popularity of space-based movies hinges on the supposed scientific plausibility of their worlds, more than an arcane debate about what counts as science fiction and what as fantasy. There is a special way in which popular culture is impactful here. People believe in Mr. Spock, that is, in extraterrestrials; they do not believe in Ariel. Extraterrestrials matter to people. Almost no one thinks mermaids are real. Mermaids are not said to be behind the 1947 UFO incident in Roswell, New Mexico; there is no mermaid-based tourism industry

A Universe of Earths. Dennis Danielson and Christopher M. Graney, Oxford University Press.
© Oxford University Press (2025). DOI: 10.1093/9780197803547.003.0018

in that city. Claims of abduction by the forces of an advanced undersea civilization of fish people are not so familiar as to be spoofed on *Saturday Night Live*.[1]

Why not? Fish people are no less scientifically plausible than extraterrestrials as an explanation for UFOs or people's strange experiences. Fish people do not require life arising on another earth through means that we can no longer explain (thanks to the demise of spontaneous generation). Fish people do not require denizens of one pale blue dot to be making vast journeys in small craft through the black immensity of space and finding another pale blue dot amid the diverse riot of countless uninhabitable worlds—only to flirt shyly with its naval aircraft or with some John and Jane Doe driving at night on a lonely two-lane road. No, the hypothesis of fish people merely requires intelligent life having evolved on *this* planet, twice—plus us being not so good at finding life in that watery part of our world (which after all far exceeds the expanse of Earth's habitable land).

Arguably, Mr. Spock is so impactful because the idea of abundant other earths and the scientific plausibility of that idea have had an impact on our popular culture and popular imagination that extends well beyond the movies or UFO true believers, reaching well into the realm of more serious media. Serious news outlets like *The New York Times* or *The Washington Post* or National Public Radio (NPR) have not mentioned mermaids as possibly being behind the UAPs (unidentified anomalous phenomena—i.e. UFOs) famously recorded by cameras on U.S. Navy aircraft. Yet they have mentioned extraterrestrials.[2] The "Report on the Historical Record of U.S. Government Involvement with Unidentified Anomalous Phenomena (UAP)" produced by the U.S. Department of Defense also makes no mention of "mermaids." But it does contain the word "extraterrestrial" more than sixty times (in a sixty-three page document)—typically in statements such as this: "AARO [All-domain Anomaly Resolution Office] assesses that the inaccurate claim that the USG [United States Government] is reverse-engineering extraterrestrial technology and is hiding it from Congress is, in large part, the result of circular reporting from a group of individuals who

---

[1] "Close Encounter—SNL," December 6, 2015, https://www.youtube.com/watch?v=PfPdYYs EfAE, celebrated in the SNL 50th Anniversary: "Close Encounter 50th—SNL50," February 16, 2025, https://www.youtube.com/watch?v=UWMN_TVDGfk. We note that indeed claims of abduction by mermaids do exist, for example in some January 2025 videos on TikTok.

[2] Cooper, Blumenthal, and Kean, "'Wow, What Is That?' Navy Pilots Reported Unexplained Flying Objects"; Simon, "We May Not Be Alone"; Rosenberg, "Former Navy pilot describes UFO encounter studied by secret Pentagon program."

believe this to be the case, despite the lack of any evidence."[3] It is no stretch to suppose that if a U.S. Department of Defense report mentioned mermaids multiple times, or spoke of mermaid technology in regard to UAPs, or if the *Times* or *Post* or NPR mentioned mermaids and U.S. Navy jets, or if serious-looking, clear-eyed people went before the U.S. Congress to speak about fish people from an advanced, undiscovered civilization built into the caverns and valleys of Earth's deep ocean floor being connected with UAPs, then they would be one vast laughing stock, and many people would surely lose their jobs.

We can only marvel at how one sort of being created by the human imagination, for whose actual existence there is no scientific evidence, is treated one way in popular culture and the media, and how another sort is treated another way. One of these sorts of beings, however, has been discussed by scientists for centuries as if it were something that anyone unhampered by a "groveling mind" would rationally believe in; the other has not. The history of ideas matters, and thus mermaid abductions are not "a thing."

And yes, this matters for science today, not just for history and media and pop culture. Those dreams of other people on kindred globes have their impact in the world of science well beyond SETI alone. In 2021 Avi Loeb, a Harvard astronomer, published *Extraterrestrial: The First Sign of Intelligent Life Beyond Earth*, with Houghton Mifflin Harcourt. In this work, he argued that the object called ʻOumuamua that passed through our Solar System in 2017 was a product of extraterrestrial technology. A serious scientist was making a serious proposal (probably unaware of the history we present here) in an attempt to explain an unusual object with an unusual trajectory.

For science, however, extraterrestrials as a ready explanation for the unexplained has its downside. In May of 2023, Dan Evans, Director of Civil Space Policy at the White House National Space Council and also a part of NASA, called to order a meeting of NASA's UAP independent study team. The team's job was to study what UAPs might be—a fair, serious question that might be tackled without presuming the existence of things for whose existence there is no evidence. But the belief that extraterrestrials *are* the answer to the UAP question was interfering with the team's work. Early in the meeting, Evans acknowledged that "public interest in UAPs is high and

---

[3] Department of Defense All-Domain Anomaly Resolution Office Report, 9.

that the demand for answers is strong."[4] Answers to what sort of questions? Some examples were later provided:[5]

- "Has NASA encountered any aliens or extraterrestrial life?"
- "What is NASA hiding and where are you hiding it?"
- "How much has been shared publicly?"
- "Has NASA ever cut the live NASA TV feed away?"

Apparently, the questions were not asked courteously. In the first few minutes of the meeting, Evans noted that some team members had been "subjected to online abuse due to their decision to participate on this panel."[6] Of course, those questions were *not* about whether NASA was hiding mermaids.

Like the Department of Defense report, the NASA UAP study panel talked about extraterrestrials. The possibility was mentioned ("however unlikely") of an "extraterrestrial source" for UAPs.[7] "Extraterrestrial civilizations" and "extraterrestrial artifacts" left in the Solar System received mention in the conversation[8]—and likewise, of course, the Big Numbers Maneuver:

> Within the scientific community there's a widespread, but by no means universal, belief that there are extraterrestrial civilizations. We have a well-developed rationale [for it that] . . . has to do with the vast numbers of exoplanets and the time scales of evolution and the possibility of convergent evolution on different planets leading to somewhat similar outcomes.[9]

But what can be done? If the NASA panel had not talked about extraterrestrials, it would have faced accusations that they were hiding things (not that there aren't such charges already).

Star Trek and *Star Wars* may be fun, but the situation just sketched is not. Free and honest discussion of science requires that we have some grasp of its history. In particular we should understand that while the Copernican model—over time, increasingly supported by science—abolished the

---

[4] "Public Meeting on Unidentified Anomalous Phenomena," at 7:04.
[5] "Public Meeting on Unidentified Anomalous Phenomena," at 3:44:15, 3:24:30.
[6] "Public Meeting on Unidentified Anomalous Phenomena," at 3:39.
[7] "Public Meeting on Unidentified Anomalous Phenomena," at 2:34:14, 2:34:51.
[8] "Public Meeting on Unidentified Anomalous Phenomena," at 2:34:00–2:36:00.
[9] "Public Meeting on Unidentified Anomalous Phenomena," at 2:35:00–2:35:39.

two-storey Universe and established Earth's status as a star, science did not support a Plurality of Worlds Universe and still doesn't. To explain 'Oumuamua or UAPs by adducing the role of extraterrestrials presupposes a Universe replete with other earths inhabited by other intelligent life. If the Universe is more sparsely inhabited—like the one civilization for every 2.5 million galaxies that Ellie's and Palmer's dialogue in *Contact* suggested— or even one civilization for every 2.5 galaxies, or even 2.5 civilizations per galaxy—then the extraterrestrials within it are going to be too far away for them ever to find Earth and send craft to pester Navy pilots, and too far away for their space junk to litter our Solar System. There simply is no scientific evidence for a Plurality of Worlds Universe. There never has been. And that matters.

So again, we marvel at how extraterrestrials have such a tight grip on our imaginations, to the point where so many humans think they are not imaginary. Why they do is a large social and psychological issue that authors more qualified than ourselves might profitably investigate. Yet we may speculate—and our speculation takes us back to the two-storey Universe.

Recall that the pre-Copernican Cosmos located us Earth-dwellers at the center—in the *lowest* place—amid a blue, unchanging, fecund, eternal Universe. Between 1543 and 1900, science forced all that aside. We came to know that we are not at the center and that the physical Universe is dark, evolving, with a beginning and a limited past, and not *full* of life. We came to know a Cosmos in which the Principle of Plenitude does not obtain. And we came to know our Earth as a pale blue dot, as a little star amid the darkness.

We seem to have thought, however, that somehow we *could* keep one bit of the old Universe. Even if Earth were a star, we could keep the "low" part if we continued to insist that we were just another star among innumerable others, that we were among the lesser intelligent beings populating the Universe, relative newcomers, with the lowest technology, occupying a low rung on the ladder of some supposed cosmic class system. Thus, we convinced ourselves that the Plurality of Worlds model was valid, and we built a whole popular and scientific culture around it.

But isn't it time to stop thinking and talking as though that Cosmos had anything to do with science? Isn't it time to abandon that last vestige of the two-storey Universe that Copernicus did away with? Yes, we may reasonably keep searching. The quest to find other earths is not irrational, not inherently unscientific or contradictory; there might be some life-bearing planets out there, and there might conceivably be some close enough to Earth that we might have at least some chance of detecting them in the foreseeable future. We must nonetheless be realistic in our expectations, and *we must convey that realism to the general public* so that no reasonable person can claim that NASA is merely hiding things.

In any case, to retain the aspersion of earthly "lowness"—to indulge in cosmological misanthropy, the denigration of our planet and its most privileged and highly developed occupants—is no mere technical error. For it resonates with a siren song tempting us to think that maybe, in the face of enormous earthly challenges, we can simply hedge our cosmic bets and become a multi-planet species—or merely stop worrying about those challenges because those other people on kindred globes are out there somewhere. Those temptations, however, must be resisted. We must accept that there is physically and practically no extraterrestrial option for the survival and flourishing of our species. There is no "Planet B." To imagine otherwise on the strength of Plurality-of-Worlds wishful thinking is thus morally and intellectually deleterious, a dangerous distraction from the scientific and practical work that our species must urgently undertake to steward, cherish, and protect planet Earth.

. . .

Behold two images of our bright blue marble surrounded by blackness (see Figure 18.1). Depending on how the photographs are cropped, the marble takes up most of the foreground, or only a tiny portion of it (the pale blue dot). Either way, most of what we know applies to this blue marble. Astronomers have indeed explored some of the mysteries of that darkness;

Figure 18.1

but inevitably it remains, for the rest of us, by comparison with Earth, boundless and ultimately impenetrable.

The dark and the bright in the picture are counterparts of the two main themes of this book. We understand how easy it would be to respond as if the immense and impenetrable dark—along perhaps with its eerie silence— were the dominant aspects of our story, but they emphatically are not, even if the dark does usefully serve to set off the bright. No, we are not Team Darkness, nor are we Team Silence either. Instead, we see ourselves as humble, devoted members of Team Earth, lovers of this planet's inhabitants, and not just the human ones. We have tried to show how the "starring" of the Earth since Copernicus offers true, scientific insight into something real, something absolutely fascinating, something worth treasuring. You might be discouraged or displeased to be told that for all *practical* purposes, physically and astronomically speaking, Earth and Earth-dwellers appear to be alone in the Universe. Yet it's true. And we appreciate that that's not a fun fact, even if there's no science—all right, no science *yet*—to show that it's not a fact.

But let's consider the claims that science does support: Earth is a planet, a wandering star, one that appears to be utterly unique and life-sustaining within our Solar System and possibly within our galactic neighborhood. Earth—again, maybe uniquely—contains complex, *intelligent* life, which includes the humans writing and reading this sentence. Science also increasingly recognizes that unless the Universe were as vast and as ancient as it is, we couldn't be here, nor could we be communicating with each other as we are doing now. The minimum takeaway here is this: It's pointless to disparage ourselves and our planet for being cosmically minuscule in size, when our very existence is cosmologically predicated on the staggering disproportions between our smallness and the inconceivable largeness of the Universe. Just ask Johannes Kepler.

We propose instead that the message of this image captured from outer space—bright blue marble against a pitch-black background—is not debilitating or discouraging. Rather, the true picture of Earth that emerges from this photograph, as well as from the history we've unfolded in this book, is exhilarating, humbling, even overwhelming. We decisively *are participants* in something genuinely special. We inhabit this picture and this history; and we matter. To ponder this message is to taste, if fleetingly, what is meant by the word *miraculous*. Like beholding the face of a newborn child and realizing the contingency and glory of a creature unique in the history of the Universe—sensing too the burden and privilege of caring for it—we hope our story reflects, albeit imperfectly, the ineffable glory of Earth, our home, our planet. In the words of David Attenborough, "It might seem like an obvious thing to say, but we need to keep saying it. Our planet is precious."[10]

And, to echo Thomas Traherne, "this little star . . . so full of mysteries" renders us awe-struck, delighted, and grateful.

[10] "Our Planet Is Precious," at 0:00.

Works Cited

"Address of Sir William Thomson, Knt., LL.D., F.R.S., President." In *Report of the Forty-First Meeting of the British Association for the Advancement of Science, held at Edinburgh in August 1871*. London: John Murray, 1872: lxxxiv–cv.

al-bīrūnī, Abu al-Rayḥān Muḥammad. *The Exhaustive Treatise on Shadows*. Edited and translated by E. S. Kennedy, 2 vols. Aleppo: Institute for the History of Science, 1976.

Al-Biruni, Abu'l-Rayhan Muhammad Ibn Ahmad. *The Book of Instruction in the Elements of the Art of Astrology*. Translated by R. Ramsay Wright. Luzac & Co., 1934.

Allegri, Alessandro. *Lettere di Ser Poi Pedante*. 1613; rpt. Casalmaggiore, 1850.

Annis, J. "Placing a Limit on Star-fed Kardashev Type III Civilisations." *Journal of the British Interplanetary Society* 52, no. 1 (1999): 33–36.

Anonymous. "To a Comet." *The Knickerbocker* 32 (1848): 214–15.

Aquinas, Thomas. *"The Summa Theologica" of St. Thomas Aquinas, Part I. QQ. I-LXXIV. Literally Translated by Fathers of the English Dominican Province, Second and Revised Edition*. Burns, Oates & Washbourne, Ltd., 1922.

Arrhenius, Svante. "Die thermophilen Bakterien und der Strahlungsdruck der Sonne." *Zeitschrift für physikalische Chemie, Stöchiometrie und Verwandtschaftslehre* 130 (1927): 516–19.

Arrhenius, Svante. *Worlds in the Making*. Translated by Dr. H. Borns. London: Harper, 1908.

Augustine (Bishop of Hippo). "The Literal Meaning of Genesis." In *The Works of Saint Augustine, a Translation for the 21st Century, Part I, Volume 13: On Genesis: A Refutation of the Manichees; Unfinished Literal Commentary on Genesis; The Literal Meaning of Genesis—introductions, translation and notes by Edmund Hill, O.P., editor John E. Rotele, O.S.A*. New City Press, 2002.

Ayala, Lucía. "Cosmology after Copernicus: Decentralisation of the Sun and the Plurality of Worlds in French Engravings." In *The Making of Copernicus: Early Modern Transformations of a Scientist and his Science*, edited by W. Neuber, C. Zittel, and T. Rahn. Brill, 2015: 201–26.

Ayala, Lucía. "On the Plurality of Worlds: Images of a New Cosmos." UC Berkeley Center for Science, Technology, Medicine & Society Lecture (May 23, 2012): https://cstms.berkeley.edu/current-events/on-the-plurality-of-worlds-images-of-a-new-cosmos/

Barrow, John, and Frank Tipler. *The Anthropic Cosmological Principle*. Oxford University Press, 1986.

Bartha, Paul. "Analogy and Analogical Reasoning." *The Stanford Encyclopedia of Philosophy*. https://plato.stanford.edu/entries/reasoning-analogy/.

Bell, Thomas. *The Jesuites Antepast*. London: William Iaggard, 1608.

Bergerac, Cyrano de. *L'Autre Monde: ou les États et Empires de la Lune.* Paris: Charles de Sercy, 1657.

Billings, L. "Alien Supercivilizations Absent from 100,000 Nearby Galaxies." *Scientific American* (April 15, 2015): https://www.scientificamerican.com/article/alien-supercivilizations-absent-from-100-000-nearby-galaxies.

Black, David C. "Worlds around Other Stars." *Scientific American* 264, no. 1 (January 1991): 76–83.

Blumenberg, Hans. *Die Genesis der Kopernikanischen Welt.* Suhrkamp, 1975.

Blumenberg, Hans. *The Genesis of the Copernican World.* Translated by Robert M. Wallace. MIT Press, 1989.

Boethius. *Consolation of Philosophy.* Translated by Philip Ridpath. London, 1785.

Bondi, Hermann. *Cosmology*, 2nd ed. Cambridge University Press, 1968.

BOTC. *The Book of the Cosmos: Imagining the Universe from Heraclitus to Hawking.* Edited by D. Danielson. 2000; Basic Books, 2001.

Brague, Remi. "Geocentrism as a Humiliation for Man." *Medieval Encounters* 3 (1997): 187–210.

Brahe, Tycho. "De Nova Stella (On a New Star, Not Previously Seen within the Memory of Any Age Since the Beginning of the World)." Translated by John H. Walden. In Harlow Shapley and Helen E. Howarth, eds., *A Source Book in Astronomy.* McGraw-Hill, 1929: 13–20.

Brück, M. T. *Agnes Mary Clerke and the Rise of Astrophysics.* Cambridge University Press, 2009.

Bruno, Giordano. *De l'infinito universo et Mondi*, 1584. Reprinted in *Le opere italiane di Giordano Bruno.* Göttingen, 1888.

Burmeister, Karl Heinz. *Georg Joachim Rhetikus, 1514–1574: Eine Bio-Bibliographie.* 3 vols. Guido Pressler Verlag, 1968.

Burton, Robert. *The Anatomy of Melancholy.* Oxford, 1621.

Calvin, John. *Commentaries on the First Book of Moses, Called Genesis, Vol. 1.* Translated by John King. Edinburgh: Calvin Translation Society, 1847.

Calvin, John. *Commentary on the Book of Psalms, Vol. 5.* Translated by James Anderson. Edinburgh: Calvin Translation Society, 1849.

Capella, Martianus. *The Marriage of Philology and Mercury*, vol. 2 in *Martianus Capella and the Seven Liberal Arts.* Translated by William Harris Stahl and Richard Johnson with E. L. Burge. Columbia University Press, 1977.

Carter, Brandon. "Large Number Coincidences and the Anthropic Principle in Cosmology." In M. S. Longair, *Confrontation of Cosmological Theories with Observational Data; Proceedings of the Symposium, Krakow, Poland, September 10–12, 1973.* Reidel, 1974.

Charalampous, Charis. "'One Common Matter' in Descartes' Physics: The Cartesian Concepts of Matter Quantities, Weight and Gravity." *Annals of Science* 76 (2019): 324–39.

Clement, Matthew S., Andre Izidoro, Sean N. Raymond, and Rogerio Deienno. "Formation of Terrestrial Planets." arXiv:2411.03453v1 (2024).

Clerke, Agnes M. *A Popular History of Astronomy during the Nineteenth Century*. Adam and Charles Black, 1908.

Clerke, Agnes Mary. *The System of the Stars*, 2nd ed. Adam and Charles Black, 1905.

Cocconi, Giuseppe, and Philip Morrison. "Searching for Interstellar Communications." *Nature* 184, no. 4690 (September 19, 1959): 844–46.

Cooper, Helene, Ralph Blumenthal, and Leslie Kean. "'Wow, What Is That?' Navy Pilots Reported Unexplained Flying Objects." *The New York Times* (May 27, 2019): A.14. https://www.nytimes.com/2019/05/26/us/politics/ufo-sightings-navy-pilots.html.

Copernicus, Nicolaus. *De revolutionibus*. Nuremberg, 1543.

Copernicus, Nicolaus. *On the Revolutions*. Translated by Edward Rosen. Polish Scientific Publications, 1978.

Cressy, David. "Early Modern Space Travel and the English Man in the Moon," *American Historical Review* 111 (2006): 961–82.

Crowe, Michael J. *The Extraterrestrial Life Debate, 1750–1900: The Plurality of Worlds from Kant to Lowell*. Cambridge University Press, 1986.

Crowe, Michael J. *The Extraterrestrial Life Debate, Antiquity to 1915: A Source Book*. University of Notre Dame Press, 2008.

Crowe, Michael J. "William Whewell, the Plurality of Worlds, and the Modern Solar System." *Zygon: Journal of Religion & Science* 51, no. 2 (2016): 431–49.

Cudworth, Ralph. *The True Intellectual System of the Universe*. 1678; London, 1820.

Danielson, Dennis. "Achilles Gasser and the Birth of Copernicanism," *Journal for the History of Astronomy* 35 (2004): 457–74.

Danielson, Dennis. "Ancestors of Apollo," *American Scientist* 99 (2011): 136–43.

Danielson, Dennis. "Scientist's Birthright: How a New Name Embodied Ideals of Connection and Inclusiveness." *Nature* 410 (2001): 1031.

Davies, Paul. *The Eerie Silence: Renewing Our Search for Alien Intelligence*. Houghton Mifflin Harcourt, 2010.

Deacon, Terrence. "Giving up the Ghost: The Epic of Spiritual Emergence." *Science & Spirit* 10, no. 2 (1997): 16–17.

Deichmann, Ute. "Origin of Life. The Role of Experiments, Basic Beliefs, and Social Authorities in the Controversies about the Spontaneous Generation of Life and the Subsequent Debates about Synthesizing Life in the Laboratory." *History and Philosophy of the Life Sciences* 34, no. 3 (2012): 341–59.

Department of Defense All-Domain Anomaly Resolution Office Report on the Historical Record of U.S. Government Involvement with Unidentified Anomalous Phenomena (UAP), Volume I (February 2024). https://media.defense.gov/2024/Mar/08/2003409233/-1/-1/0/DOPSR-CLEARED-508-COMPLIANT-HRRV1-08-MAR-2024-FINAL.PDF

Dick, Steven J. *Space, Time, and Aliens: Collected Works on Cosmos and Culture*. Springer, 2020.

Dick, Steven. *Life on Other Worlds: The 20th-Century Extraterrestrial Life Debate*. Cambridge University Press, 1998.

Dicke, R. H. "Dirac's Cosmology and Mach's Principle." *Nature* 192, no. 4801 (November 4, 1961): 440–41.

Digges, Thomas. *Alae seu scalae mathematicae.* London, 1573.

Digges, Thomas. *A Perfit Description of the Caelestiall Orbes.* London, 1576.

Diodorus of Sicily. "The Library of History." In *Diodorus of Sicily in Twelve Volumes.* Translated by C. H. Oldfather. Cambridge, Mass: Harvard University Press, 1933.

Dobbell, Clifford, F.R.S. *Antony van Leeuwenhoek and His "Little Animals."* Harcourt Brace, 1932.

Dobzhansky, Theodosius. *Man's Place in the Universe: Changing Concepts.* Edited by David W. Corson. University of Arizona College of Liberal Arts, 1977.

Donne, John. *The Complete English Poems.* Edited by A. J. Smith. Penguin, 1971.

Dorminey, Bruce. "The Search for Astroengineers." *Physics World* 21, no. 4 (April 2008): 32.

Drake, Frank. "A Reminiscence of Project Ozma." *Cosmic Search* 1, no. 1 (January 1979): 10–15. http://www.bigear.org/CSMO/HTML/CSIntro.htm; http://www.bigear.org/cosmic_search.htm.

Drake, Frank, and Dava Sobel. *Is Anyone Out There?: The Scientific Search for Extraterrestrial Intelligence.* Dell, 1994.

Dreyer, J. L. E. *Tycho Brahe: A Picture of Scientific Life and Work in the Sixteenth Century.* Edinburgh, 1890.

Dreyer, J. L. E. ed. *Tychonis Brahe Dani: Epistolæ Astronomicæ I [Tychonis Brahe Dani Opera Omnia VI].* Nielsen & Lydiche, 1919.

Eddington, Arthur. "The End of the World: from the Standpoint of Mathematical Physics." *Nature* 127 (1931): 447–53.

Edgerton, Samuel Y., Jr. *The Heritage of Giotto's Geometry: Art and Science on the Eve of the Scientific Revolution.* Cornell University Press, 1991.

Farago, Claire, Janis Bell, and Carlo Vecce, eds. *The Fabrication of Leonardo da Vinci's "Trattato della pittura."* Brill, 2018.

Farley, John. "The Spontaneous Generation Controversy (1859–1880): British and German Reactions to the Problem of Abiogenesis." *Journal of the History of Biology* 5, no. 2 (1972): 285–319.

Fontenelle, Bernard Le Bovier de. *A Conversation on the Plurality of Worlds: Translated from the French of M. De Fontenelle.* London, 1758.

Frank, Adam. *The Little Book of Aliens.* HarperCollins, 2023.

"Frank Drake, Who Led Search for Life on Other Planets, Dies at 92." *The New York Times,* September 7, 2022: https://www.nytimes.com/2022/09/05/science/space/frank-drake-dead.html.

Freud, Sigmund. *A General Introduction to Psycho-Analysis.* Liveright Publishing, 1935.

Gagarin, Yuri. *Road to the Stars* [1962]. University Press of the Pacific, 2002.

Galilei, Galileo. *Dialogue Concerning the Two Chief World Systems: Ptolemaic and Copernican.* Translated and with revised notes by Stillman Drake. Modern Library-Random House, 2001.

Galilei, Galileo, and Christoph Scheiner. *On Sunspots.* Edited and translated by Eileen Reeves and Albert Van Helden. University of Chicago Press, 2010.

Godwin, Francis. *The Man in the Moon: or a Discourse of a Voyage Thither.* London, 1638; Reprint: William Poole, ed. Broadview, 2009.

Goethe, Johann Wolfgang. "Materialien zur Geschichte der Farbenlehre." In Erich Trunz *Goethes Werke*, Hamburger Ausgabe, vol. 14. Christian Wegner Verlag, 1960.

Graney, Christopher M. "Galileo Between Jesuits: The Fault Is in the Stars." *The Catholic Historical Review* 107, no. 2 (2021): 191–225.

Graney, Christopher M. *Setting Aside All Authority: The Science Against Copernicus in the Age of Galileo*. University of Notre Dame Press, 2015.

Graney, Christopher M. "The Starry Universe of Jacques Cassini: Century-old Echoes of Kepler." *Journal for the History of Astronomy* 52, no. 2 (2021): 147–67.

"Greetings to Winners of Stalin Prizes." *Pravda* (March 5, 1951): 4.

Griffith, Roger L., Jason T. Wright, Jessica Maldonado, Matthew S. Povich, Steinn Sigurðsson, and Brendan Mullan. "The Ĝ Infrared Search for Extraterrestrial Civilizations with Large Energy Supplies. III. The Reddest Extended Sources in *WISE*." *The Astrophysical Journal Supplement Series* 217, no. 2 (2015): 1–34.

Guthke, Karl S. *The Last Frontier: Imagining Other Worlds, from the Copernican Revolution to Modern Science Fiction*. Translated by Helen Atkins. Cornell University Press, 1990.

Hamel, Jürgen. *Nicolaus Copernicus: Leben, Werk und Wirkung*. Spektrum Verlag, 1994.

Hart, Michael H. "An Explanation for the Absence of Extraterrestrials on Earth." *Quarterly Journal of the Royal Astronomical Society* 16 (June 1975): 128–35.

Hawking, Stephen, and George Ellis. *The Large-Scale Structure of Space-Time*. Cambridge University Press, 1973.

Helmholtz, Hermann von. "Vorrede." In W. Thomson and P. G. Tait, *Handbuch der Theoretischen Physik, Autorisierte Deutsche Übersetzung von Dr. H. Helmholtz und G. Wertheim*, 1. Band, 2. Theil. Braunschweig, 1874: v–xiv.

Hernandez, Joe. "William Shatner Boldly Went into Space for Real. Here's What He Saw." NPR: Space (October 13, 2021). https://www.npr.org/2021/10/13/1045377132/william-shatner-star-trek-captain-kirk-blue-origin-space-flight.

Herschel, John. "Letter to W. Whewell of 3 January 1854." In Crowe, *The Extraterrestrial Life Debate: Antiquity to 1915*: 358–60.

Herschel, John F. W. *A Treatise on Astronomy*. Philadelphia, 1838.

Herschel, William. "On the Nature and Construction of the Sun and the Fixed Stars." In *The Philosophical Transactions of the Royal Society of London for the Year MDCCXCV, Part I*. London, 1795: 46–72.

Hill, John. *Urania, or a Compleat View of the Heavens Containing the Antient and Modern Astronomy in Form of a Dictionary*, vol. 1. London, 1754.

Hoeppe, Götz. *Why the Sky Is Blue: Discovering the Color of Life*. Translated by John Stewart. Princeton University Press, 2007.

Horrocks, Jeremiah. *Venus Seen on the Sun: The First Observation of a Transit of Venus*. Translated by Wilbur Applebaum. Brill, 2012.

Hubble, Edwin. "The Problem of the Expanding Universe." *Science* 95 (1942): 212–15.

Huygens, Christiaan. *Celestial Worlds Discover'd*, 2nd ed. London, 1722.

Kamminga, Harmke. "Life from Space—A History of Panspermia." *Vistas in Astronomy* 26 (1982): 67–86.

Kant, Immanuel. *Kant's Cosmogony*. Translated by W. Hastie. Glasgow, 1900.

Kant, Immanuel. *Universal Natural History and Theory of the Heavens*. Translated by Ian Johnston. https://web.viu.ca/johnstoi/kant/kant2e.htm

KCGSM. Kepler, Johannes. *Kepler's Conversation with Galileo's Sidereal Messenger*. Translated by Edward Rosen. Johnson Reprint Corporation, 1965.

Kepler, Johannes. "Chapter 16 of *De Stella Nova*." Translated by Christopher M. Graney. In *Kepler's New Star (1604): Context and Controversy*, ed. Patrick J. Boner. Brill, 2021: 48–59.

Kepler, Johannes. "Chapter 21 of *De Stella Nova*." Translated and abridged by Edward Rosen, in Kepler, *Kepler's Conversation with Galileo's Sidereal Messenger*: 127–29.

Kepler, Johannes. "*Dioptrics*." In Aviva Rothman, ed. and trans., *The Dawn of Modern Cosmology: From Copernicus to Newton*. Penguin Random House, 2023: 203–15.

Kepler, Johannes. *Epitome of Copernican Astronomy and Harmonies of the World*. Translated by C. G. Wallis. Prometheus Books, 1995.

Kepler, Johannes. *Somnium: The Dream*. Translated by Edward Rosen. University of Wisconsin Press, 1967.

King, Bob. "Solving an Earthshine Mystery." *Sky & Telescope* website (July 27, 2022): https://skyandtelescope.org/observing/earthshine-mystery.

Lalande, Jérôme de. "Introductory Note." In Bernard de Fontenelle, *Conversations on the Plurality of Worlds*, trans. Elizabeth Gunning London, 1803: iii–ix.

Lane, Nick. *The Vital Question: Energy, Evolution, and the Origins of Complex Life*. Norton, 2015.

Leeuwenhoek, Antony Van. "On the Generation of Eels." In *Select Works of Antony Van Leeuwenhoek*, trans. S. Hoole, vol. 2. London, 1816: 62–64.

Leiman, Sid Z. "R. Israel Lipshutz and the Mouse That Is Half Flesh and Half Earth: A Note on Torah U-Madda in the Nineteenth Century." In Yaakov Elman and Jeffrey S. Gurock, eds., *Ḥazon Naḥum: Studies in Jewish Law, Thought, and History Presented to Dr. Norman Lamm on the Occasion of his Seventieth Birthday*. Yeshiva University Press, 1997: 449–58.

Leonhard, Karin. "Pictura's Fertile Field: Otto Marseus van Schrieck and the Genre of Sottobosco Painting." *Simiolus: Netherlands Quarterly for the History of Art* 34, no. 2 (2009/2010): 95–118.

LePage, Andrew J. "Where They Could Hide." *Scientific American* 283 (July 2000): 40–41.

Lewis, C. S. *The Discarded Image: An Introduction to Medieval and Renaissance Literature*. 1964; reprinted Cambridge University Press, 2012.

Lipínski, E. "Shemesh." In *Dictionary of Deities and Demons in the Bible*. Edited by Karel van der Toom et al., 2nd ed. Brill, 1999: 822–23.

Livio, Mario. *Galileo and the Science Deniers*. Simon and Schuster, 2020.

Lovejoy, A. O. *The Great Chain of Being: A Study of the History of an Idea*. 1936: rpt. Harper, 1960.

Lucretius. *De Rerum Natura*. Translated by Thomas Creech, 2nd ed. London, 1683.

Machamer, Peter, ed. *Cambridge Companion to Galileo*. Cambridge University Press, 1998.

Magnus, Albertus. *Commentarii in II Sententiarum*.

Mason, Stephen F. *A History of the Sciences*. Macmillan, 1962.

McMullin, Ernan. "Bruno and Copernicus." *Isis* 78 (1987): 55–74.

MathDisq: Johann Georg Locher with Christoph Scheiner. *Disquisitiones Mathematicae*. Ingolstadt, 1614; translated in Christopher M. Graney. *Mathematical Disquisitions: The Booklet of Theses Immortalized by Galileo*. University of Notre Dame Press, 2017.

Miller, Stuart S. "'Epigraphical' Rabbis, Helios, and Psalm 19." *The Jewish Quarterly Review*, New Series, 94, no. 1 (2004): 27–76.

Montaigne, Michel de. "An Apology of Raymond Sebond." In *The Essays of Michel de Montaigne*, translated by Charles Cotton. London, 1892.

More, Henry. *Democritus Platonissans, or, An Essay upon Infinity of Worlds out of Platonick Principles*. Cambridge, 1646.

Nadis, Steve, and Shing-Tung Yau. *The Gravity of Math: How Geometry Rules the Universe*. Basic Books, 2024.

Newton, Isaac. *Mathematical Principles of Natural Philosophy, translated into English*, Vol. 2. London, 1729.

Nicolson, Marjorie Hope, ed. *Conway Letters: The Correspondence of Anne, Viscountess Conway, Henry More, and their Friends, 1642–1684*. Oxford University Press, 1930.

"Our Planet Is Precious: Reasons for Hope." Cambridge University [video] (November 14, 2024). https://www.youtube.com/watch?v=uIfiOKQ67CI.

Palingenius, Marcellus. *The Zodiake of Life*. Translated by Barnaby Googe. London, 1565.

Pannekoek, Antonie. *A History of Astronomy*. Interscience Publishers, 1961.

Parsons, Robert. *A Review of Ten Publike Disputations*. [np] 1604.

Pascal, Blaise. *Pensées* [ca. 1650]. From *Thoughts*, translated by W. F. Trotter. Collier, 1910.

Pico, Giovanni. *Opera Omnia Ioannis Pici*. Basel, 1493.

Pico, Giovanni. "Oration on the Dignity of Man." In *The Renaissance Philosophy of Man*. Edited by Ernst Cassirer et al. University of Chicago Press, 1948: 223–55.

Pliny the Elder. *The Historie of the World*. Translated by Philemon Holland. London, 1601.

Poe, Edgar A. *Eureka: A Prose Poem*. New York, 1848.

Polanyi, Michael. *Personal Knowledge*. University of Chicago Press, 1958.

Poole, Robert. *Earthrise: A Short History of the Whole Earth*, 2nd ed. 1957Books, 2023.

Prowe, Leopold. *Nicolaus Coppernicus*, 2 vols. 1883–1884; rpt. Zeller-Verlag, 1967.

Ptolemy. "Almagest." In *Great Books of The Western World (16): Ptolemy, Copernicus, Kepler*. W. Benton, 1952.

"Public Meeting on Unidentified Anomalous Phenomena (Official NASA Broadcast)." NASA Video (May 31, 2023). https://www.youtube.com/watch?v=bQo08JRY0iM.

Rankine, William John Macquorn. "On the Reconcentration of the Mechanical Energy of the Universe." In *The London, Edinburgh and Dublin Philosophical Magazine and Journal of Science, Volume 4, Fourth Series*. London, 1852: 358–60.

Rees, Martin. *Before the Beginning: Our Universe and Others*. Addison-Wesley/Helix Books, 1998.

Reinhold, Timo, et al. "The Sun Is Less Active than Other Solar-like Stars." *Science* 368 (2020): 518–21.

"Review and Outlook—Mars." *Wall Street Journal* (December 28, 1907): Morning Edition, front page.

Rheticus, Georg Joachim. *Narratio Prima [First Account]*. Gdańsk, 1540.

Roberts, Isaac. "Photographs of the Nebulæ M31, h44, and h51 Andromedæ, and M27 Vulpeculæ." *Monthly Notices of the Royal Astronomical Society* 49 (1888): 65.

Romer, Alfred. "The Welcoming of Copernicus's *De revolutionibus*: The *Commentariolus* and its Reception." *Physics in Perspective* 1 (1999): 157–84.

Rosen, Edward, trans. *Three Copernican Treatises*, 3rd ed. Octagon Books, 1971.

Rosenberg, Eli. "Former Navy Pilot Describes UFO Encounter Studied by Secret Pentagon Program." *The Washington Post* (December 18, 2017). https://www.washingtonpost.com/news/checkpoint/wp/2017/12/18/former-navy-pilot-describes-encounter-with-ufo-studied-by-secret-pentagon-program/.

Rossi, Paolo. "Nobility of Man and Plurality of Worlds." In *Science, Medicine and Society in the Renaissance*, ed. Allen G. Debus. Science History Publications, 1972: 2:131–62.

Ruestow, Edward G. "Leeuwenhoek and the Campaign against Spontaneous Generation." *Journal of the History of Biology* 17, no. 2 (1984): 225–48.

Sagan, Carl. *The Demon Haunted World: Science as a Candle in the Dark*. Ballantine Books, 1997.

Sagan, Carl. *Pale Blue Dot: A Vision of the Human Future in Space*. Random House, 1994.

Sagan, Carl. "Quest for Extraterrestrial Intelligence." *Cosmic Search* 1, no. 2 (March 1979): 2–8.http://www.bigear.org/CSMO/HTML/CSIntro.htm; http://www.bigear.org/cosmic_search.htm.

Salisbury, Harrison E. "Russian Astronomers Hold Theory of Cosmos Origin Surpasses West." *The New York Times* (July 14, 1949): 1.

Scharf, Caleb. *The Copernicus Complex: The Quest for Our Cosmic (In)Significance*. Allen Lane, 2014.

Shackleford, Joel. "Myth 7. That Giordano Bruno Was the First Martyr of Modern Science." In *Galileo Goes to Jail and Other Myths about Science and Religion*, ed. Ronald L. Numbers. Harvard University Press, 2009: 59–67.

Shapley, Harlow. *The View from a Distant Star*. Basic Books, 1963.

Simon, Scott. "We May Not Be Alone." NPR *Weekend Edition Saturday* (December 23, 2017). https://www.npr.org/2017/12/23/573013655/we-may-not-be-alone.

Smith, Howard. "Alone in the Universe." *Zygon* 51, no. 2 (June 2016): 497–519.

Smith, Howard A. "Questioning Copernican Mediocrity." *American Scientist* 105, no. 4 (July-August 2017): 232–39.

Snobelen, Stephen D. "Apocalyptic Themes in Isaac Newton's Astronomical Physics." In A. Ricker, C. J. Corbally, and D. Dinnell, eds., *Intersections of Religion and Astronomy*. Routledge, 2021: 95–104.

Spies, Otto. "Al-Kindī's Treatise on the Cause of the Blue Colour of the Sky." *Journal of the Bombay Branch of the Royal Asiatic Society*, n.s., 13 (1937): 7–19.

Stebbing, L. S. *A Modern Introduction to Logic*. Methuen, 1933.

Swerdlow, Noel M. "The Derivation and First Draft of Copernicus' Planetary Theory: A Translation of the *Commentariolus* with Commentary." *Proceedings of the American Philosophical Society* 117 (1973): 423–512.

Tanner, John. "'And Every Star Perhaps a World of Destined Habitation': Milton and Moonmen." *Extrapolation* 30 (1989): 267–79.

Thomson, William. "On Mechanical Antecedents of Motion, Heat, and Light." In *Report of The British Association for the Advancement of Science: Twenty-Fourth Meeting; Held at Liverpool in September 1854.* London, 1855; addendum "Notices and Abstracts of Miscellaneous Communications to the Sections": 59–63.

Tolosani, Giovanni Maria. "Tolosani's Condemnations of Copernicus' *Revolutions.*" In Edward Rosen, ed., *Copernicus and the Scientific Revolution.* Krieger, 1984: 188–91.

Topsell, Edward. *The History of Four-Footed Beasts and Serpents.* London, 1658.

Traherne, Thomas. *Poetry and Prose.* Edited by Denise Inge. SPCK, 2002.

Tropp, E. A., V. Y. Frenkel, and A. D. Chernin. *Alexander A. Friedmann: The Man Who Made the Universe Expand.* Translated by A. Dron and M. Burov. Cambridge University Press, 1993.

Van Helden, Albert. *Measuring the Universe: Cosmic Dimensions from Aristarchus to Halley.* University of Chicago Press, 1985.

von Guericke, Otto. *The New (So-Called) Magdeburg Experiments of Otto von Guericke.* Translated by Margaret Glover Foley Ames. Springer, 1994.

Wallace, Alfred R. *Man's Place in the Universe.* Chapman and Hall, Ltd., 1904.

Ward, Peter Douglas, and Donald Brownlee, *Rare Earth: Why Complex Life Is Uncommon in the Universe.* Springer-Verlag/Copernicus, 2000.

Whewell, William (anonymous). *Of the Plurality of Worlds: An Essay.* London, 1853.

Wickramasinghe, Chandra. "Panspermia According to Hoyle." *Astrophysics and Space Science* 285 (2003): 535–38.

Wilkins, John. *Discourse Concerning a New Planet, Tending to Prove, That 'tis Probable Our Earth Is One of the Planets.* London, 1640.

Wilkins, John. *The Discovery of a World in the Moone: or, A Discourse Tending to Prove That 'tis Probable There May Be Another Habitable World in That Planet.* London, 1638.

Wootton, David. *The Invention of Science.* Allen Layne/Penguin Random House, 2015.

Acknowledgments

A full accounting of our debts of gratitude would demand many pages and span many years stretching across two careers and almost two lifetimes. For specific purposes of this happily coauthored book, however, we offer warm thanks to friends and family members who have read, commented on, and critiqued earlier portions or versions of our work: Ned Berghausen, Jessa Danielson, Janet Henshaw Danielson, Christina Graney, and Virginia Trimble. All remaining flaws and infelicities we acknowledge as our own.

Index

For the benefit of digital users, indexed terms that span two pages (e.g., 52–53) may, on occasion, appear on only one of those pages.

Note: Figures are indicated by an italic *f*.

The manufacturer's authorised representative in the EU for product safety is Oxford University Press España S.A. of El Parque Empresarial San Fernando de Henares, Avenida de Castilla, 2 – 28830 Madrid (www.oup.es/en or product.safety@oup.com). OUP España S.A. also acts as importer into Spain of products made by the manufacturer.

Printed in the USA/Agawam, MA
January 22, 2026

900178.129